Non-Traditional Military Training
for Canadian Peacekeepers

Non-Traditional Military Training for Canadian Peacekeepers

a study prepared for

the Commission

of Inquiry into

the Deployment of

Canadian Forces

to Somalia

Paul LaRose-Edwards
Jack Dangerfield
Randy Weekes

© Minister of Public Works and Government Services Canada 1997
Printed and bound in Canada

Available in Canada through
your local bookseller or by mail from
Public Works and Government Services Canada — Publishing
Ottawa, Canada K1A 0S9

Catalogue No. CP32-64/4-1997E
ISBN 0-660-16881-2

Canadian Cataloguing in Publication Data

LaRose-Edwards, Paul

Non-traditional military training for Canadian peacekeepers

Issued also in French under title: Instruction militaire non traditionnelle destinée aux casques bleu canadiens.
ISBN 0-660-16881-2
Cat. no. CP32-64/4-1997E

1. Military education — Canada.
2. Military missions — Canada.
I. Dangerfield, Jack.
II. Weekes, Randy.
III. University of Ottawa. Human Rights Research and Education Centre.
IV. Commission of Inquiry into the Deployment of Canadian Armed Forces to Somalia.
V. Title.

U440.L37 1997   355.5'0971   C97-980067-6

Contents

ABBREVIATIONS   ix

PREFACE   xi

RECOMMENDATIONS   xiii

CHAPTER ONE — INTRODUCTION: RATIONALE AND POLICY   1
The Rationale for Non-Traditional Military Training for Peacekeeping   1

Canadian Government Policy   6
   Parliament   6
   Department of National Defence   7

United Nations   8

Imminent Change   9

CHAPTER TWO — TRAINING IN THE CANADIAN MILITARY   11
Canadian Forces Training Framework   11
   Individual Training   12
   Collective Training   14
   Canadian Forces Training Principles   14

Background for Canadian Forces Peacekeeping Training   15
   NDHQ Program Evaluation E2/90   16
   DCDS Directive 4500-1 (DCDS) 29 December 1993   18

vi   Contents

  DCDS Study Directive on Peacekeeping Training in the
  Canadian Forces   22
  UN Peacekeeping Training Assistance Teams   22
  Doctrine Publications   23

 Interviews   27
  National Defence Headquarters   28
  Land Force Command Headquarters   31
  Land Force Central Area Headquarters   32
  Headquarters 2 Canadian Brigade Group   32
  1st Battalion the Royal Canadian Regiment and the Royal
  Canadian Dragoons   34
  Canadian Forces Command and Staff College   37
  Canadian Land Force Command and Staff College   37
  Army Lessons Learned Centre and Peace Support
  Training Centre   39
  Royal Military College   40
  Canadian Forces Recruit School   40
  The Royal Canadian Regiment Battle School   41
  Office of the Judge Advocate General   41

 Study Team General Observations   42

CHAPTER THREE — RECOMMENDATIONS ON TRAINING REQUIREMENTS   45
Introduction   45

What Skills   47
 Skills for UN Peacekeeping   47
 Strategic   49
 Operational   50
 Theatre Environment   51
 Specialized Training   53
 What Other Peacekeeping Skills   58

Training: For Whom, When, Where   59
 General   59
 National Defence Headquarters   62
 Command Headquarters   64
 Land Force Area Headquarters   65

Headquarters Brigade Groups, and Similar Sea and Air Entities 66
Battalions, Regiments, Air Squadrons, and Other Similar Size Units 66
Canadian Forces Command and Staff College, and Canadian Land Force Command and Staff College 68
Army Lessons Learned Centre and Proposed Peace Support Training Centre 68
Royal Military College 70
Canadian Forces Recruit School and Other Basic Training Establishments 71
Battle Schools 71
Reserves 72
Office of the Judge Advocate General 73

Other Sources of Training Practice and Guidance 75
    Pearson Peacekeeping Centre 75
    Canadian Civilian Police 76
    Auditor General 78
    Other Countries' Military Training for Peacekeeping 79
    United Nations 80

CHAPTER FOUR — CONCLUSION 83

NOTES 85

APPENDIX I DCDS Directive 4500-1 (DCDS) 29 December 1993 Training Requirements for Peacekeeping Operation 93

APPENDIX II DCDS Study Directive 4500-1 (DCDS) 14 September 1995 Peacekeeping Training in the Canadian Forces 111

Abbreviations

ALLC	Army Lessons Learned Centre
CDS	Chief of Defence Staff
CF	Canadian Forces
CFCSC	Canadian Forces Command and Staff College
CFLCSC	Canadian Forces Land Command and Staff College
CFTS	Canadian Forces Training System
CIDA	Canadian International Development Agency
CIVPOL	UN Civilian Police
CMOC	Civilian-Military Operations Centre
DCDS	Deputy Chief of Defence Staff
DFAIT	Department of Foreign Affairs and International Trade
DHA	UN Department of Humanitarian Affairs
DND	Deparment of National Defence
DPA	UN Department of Political Affairs
DPKO	UN Department of Peacekeeping Operations
GPCT	General Purpose Combat Training
IBTS	Individual Battle Task Standards
ICRC	International Commission of the Red Cross
JTF	Joint Task Force
LFC	Land Force Command
MSF	Médecins Sans Frontières
MTAP	Military Training Assistance Program
NCM	Non Commissioned Member
NDHQ	National Defence Headquarters
NGO	Non-Governmental Organization
OPDP	Officer Professional Development Program
PKO	Peacekeeping Operation
PPC	Pearson Peacekeeping Centre

RMC	Royal Military College
SOFA	Status of Forces Agreement
SOP	Standing Operating Procedure
TAT	UN Training Assistance Team
TCN	Troop Contributing Nation
UNDP	United Nations Development Program
UNHCR	(Office of the) United Nations High Commissioner for Refugees
UNITAR	United Nations Institute for Training and Research
VCDS	Vice Chief of Defence Staff
WFP	World Food Program

Preface

The Human Rights Research and Education Centre at the University of Ottawa was commissioned by the Somalia Inquiry to produce this study and recommendations on non-traditional military training for Canadian military in preparation for peacekeeping.

The research team consists of Paul LaRose-Edwards (team leader), LGen. (ret'd) Jack Dangerfield, and Randy Weekes.

The study proceeds from the increasing recognition that traditional military training, while critical for successful peacekeeping, is not in itself sufficient in present-day peacekeeping operations or complex emergencies. From that point of departure and in pursuance of one of the mandates of the Commission, Chapter 1 sets out the rationale for non-traditional military training for modern peacekeeping operations.

Chapter 2 then identifies how those requirements have or have not been met for Canadian units and individuals deploying on UN peacekeeping missions, and describes both training received as part of the Department of National Defence's regular training cycles, and mission-specific predeployment training. An important corollary will be the identification of steps being taken by the Canadian military to address certain deficiencies.

Finally and most critically, Chapter 3 of the study identifies the additional training needs for the Department of National Defence with concrete recommendations on when, where, and how such needs could be met.

Paul LaRose-Edwards
LGen. (ret'd) Jack Dangerfield
Randy Weekes

Ottawa, 19 December 1995

Recommendations

RECOMMENDATION 1: It is recommended that the Canadian Forces overall training philosophy be amended so that general-purpose combat training, while remaining the foundation of training policy, is supplemented by additional non-traditional military training geared specifically for UN peacekeeping operations. p. 46

SKILLS FOR UN PEACEKEEPING

Strategic

RECOMMENDATION 2: It is recommended that military and civilian personnel selected for positions involving peacekeeping operations receive training (at the strategic level) on subjects such as UN decision-making, mandate formulation and interpretation, UN and national command and control mechanisms, and rules of engagement formulation and interpretation. p. 49

RECOMMENDATION 3: It is recommended that doctrine be developed on the concept of "unity of effort" in UN operations, e.g., operating within the normally loose and poorly defined UN chains of command which frequently involve civilian organizations, and that this doctrine be practised by the Canadian Forces during some of their collective training exercises. p. 50

Operational

RECOMMENDATION 4: It is recommended that Canadian military receive training on the unique character of UN operations in such areas as its standing operating procedures, administration, logistics, and terminology. p. 50

RECOMMENDATION 5: It is recommended that Canadian military receive training on dealing with other military and civilian field partners, so as to increase Canadian ability to play a role in enhancing unity of effort by all civilian-military components of a UN field operation. p. 51

Theatre Environment

RECOMMENDATION 6: It is recommended that a guide to the mission's cultural behaviour context, including factors such as religion where significant, be prepared centrally and distributed to all individuals or unit members deploying on mission. This should be carried out by a central responsibility centre which is also tasked with collecting and articulating lessons learned for subsequent guides and troop rotations. p. 52

RECOMMENDATION 7: It is recommended that a training session on dealing with the local population, involving nationals from the mission area or subject matter experts, be an element of each unit's pre-deployment preparation, and that guidance and sourcing support for such training be provided by a central responsibility centre. p. 52

RECOMMENDATION 8: It is recommended that, as much as possible, subjects such as country briefs, population details, ethnic characteristics, culture, etc., be largely taught by experts or unit officers rather than by intelligence cells. p. 52

RECOMMENDATION 9: It is recommended that at least one individual per battalion-size unit deployed on peacekeeping be sufficiently trained to speak the predominant local language(s), and that other peacekeepers using translators be trained on their capacities and limitations. Support for further self-directed language learning should also be provided in the field. p. 53

Specialized Training

RECOMMENDATION 10: It is recommended that low-level conflict mediation be taught to all junior NCMs, and that more refined mediation and conflict resolution skills be taught to senior NCMs and officers. This training should largely occur as part of regular professional and unit training, but should be customized during pre-deployment refresher training to address the particular cultural/political environment of the theatre of operations. p. 54

RECOMMENDATION 11: It is recommended that the Canadian Forces train various military officers as specialists in human rights monitoring and reporting, both to work with those UN field staff coordinating human rights promotion and protection, and also to interpret human rights intelligence to guide Canadian peacekeeping tactical decisions. p. 55

RECOMMENDATION 12: It is recommended that DND train various military officers as specialists in humanitarian assistance, both to facilitate military field support for the traditional agencies providing such assistance, and also to advise any Canadian peacekeeping units that might be specifically tasked to provide humanitarian assistance. p. 56

RECOMMENDATION 13: It is recommended that the Canadian Forces train various military officers, particularly those with engineer and support roles, as specialists in post-conflict rehabilitation so as to maximize the contribution of the training, skills and equipment of certain peacekeeping units or sub-units such as field engineers that might be present yet relatively underutilized during various stages of a peacekeeping mission. Similarly, such specialists could advise units that are specifically deployed to effect post-conflict rehabilitation. p. 56

RECOMMENDATION 14: It is recommended that critical incident stress management be emphasized as a key component of general combat readiness, and that training to manage critical incident stress be augmented to deal with incidents relatively unique to peacekeeping operations. p. 57

RECOMMENDATION 15: It is recommended that teaching the law of armed conflict become much more prevalent and extensive, and that it be taught as an operations subject with clear field applicability as opposed to a legal skill. It is important that international human rights law and standards, particularly as refined by the UN for low-level conflict CIVPOL functions, be incorporated into such training. p. 58

What Other Peacekeeping Skills

RECOMMENDATION 16: It is recommended that J3 Peacekeeping, as the office of primary interest, create and chair a DND-wide working group to undertake the identification in depth and in detail of non-traditional military skills needed for peacekeeping. p. 58

xvi Recommendations

TRAINING: FOR WHOM, WHEN, WHERE

General

RECOMMENDATION 17: It is recommended that once DND has identified in greater detail the content of nontraditional military training for peacekeeping, that J3 Peacekeeping, as the office of primary interest, create and chair a working group which would include the Director of Military Training and Education (Directorate of Military Personnel), as a key office of collateral interest, to undertake the identification of which components of DND, officers, senior NCMs, all NCMs, and civilians need to receive non-traditional military training for peacekeeping. p. 59

RECOMMENDATION 18: It is recommended that the Canadian Forces develop a core program of nontraditional training that will be received by all components of the Forces, and those civilians of DND who are involved in these operations. p. 60

RECOMMENDATION 19: It is recommended that the pre-deployment training period should be at least 90 days. This may be reduced if the unit was on UN standby and may need to be increased if the unit is composite or has a lot of augmentees. Training for individuals is more a variable depending on the mission, but needs to be extended beyond the few days now spent on this training to a period of about 14 to 21 days (more for observers, less for staff officers). p. 61

RECOMMENDATION 20: It is recommended that in light of the finite scope for the pre-deployment training period and the limits that imposes on non-traditional and mission-specific training, a core of peacekeeping subjects be taught in advance at regular stages in unit and individual training. These peacekeeping skills, as with general combat readiness, will be merely refreshed and refined during the annual and pre-deployment periods. p. 61

RECOMMENDATION 21: It is recommended that non-traditional military peacekeeping training be an integral part of most existing military training mechanisms and establishments. p. 62

xvii Recommendations

National Defence Headquarters

RECOMMENDATION 22: It is recommended that NDHQ make it clear in both the wording and the spirit of training policy that Canadian peacekeepers in the 1990s require enhanced nontraditional military training for peacekeeping. p. 62

RECOMMENDATION 23: It is recommended that NDHQ create a single, central and joint peacekeeping training section within its organization with primacy amongst the staff matrix. This section would work closely with the soon-to-be-created Peace Support Training Centre to be set up under Land Force Command, but would not be replaced by it. p. 62

RECOMMENDATION 24: It is recommended that NDHQ examine methods of quickly disseminating operational information needed by units about to deploy, so as to allow them to effectively design and deliver their unit training. p. 63

RECOMMENDATION 25: It is recommended that the Officer Professional Development Council examine the mandates given to CF staff colleges, military colleges and personnel sections with a view to formalizing peacekeeping training objectives for the various levels of an officer's professional development. p. 63

RECOMMENDATION 26: It is recommended that the policy of not having direct contact with in-place units be examined with a view to allowing replacement units to have contact with in-place units for training matters. p. 64

Command Headquarters

RECOMMENDATION 27: It is recommended that the Commands institutionalize a flowing and coherent system of analysis of peacekeeping policy, the originating of peacekeeping doctrine (initially as a single service but inputting into joint, tri-service doctrine), and the creation of peacekeeping training standards. p. 64

RECOMMENDATION 28: It is recommended that the evolution of the Army Lessons Learned Centre and the creation of a Peace Support Training Centre at Land Force Command be pursued with vigour and that these centres be tied into the above system to provide both a corporate memory based on past experiences, and an input into future doctrine production. It is emphasized that the resourcing of these centres should not be at the expense of a national tri-service focal point. p. 64

RECOMMENDATION 29: It is recommended that, to consciously protect the time allocated to peacekeeping training, the Commands examine notionally splitting pre-deployment training into three overlapping blocks: general-purpose combat training; peacekeeping training; and departure administration. p. 65

RECOMMENDATION 30: It is recommended that the Commands review the mandates given to their staff colleges, warfare schools and similar institutions, with a view to enhancing the peacekeeping training objectives of those institutions. p. 65

Land Force Area Headquarters

RECOMMENDATION 31: It is recommended that all Land Force Area Headquarters assume full responsibility for training and screening all augmentees so that they arrive at a deploying unit at the same level of general-purpose combat capability (battle task standard) as the personnel of the deploying unit. p. 65

RECOMMENDATION 32: It is recommended that Land Force Area Headquarters be the principal interface with non-military organizations, and be the channel for providing training assistance from those organizations (e.g., civilian police, Red Cross, Canadian peacekeeping partners, etc.). p. 66

Headquarters Brigade Groups, and Similar Sea and Air Entities

RECOMMENDATION 33: It is recommended, particularly for a first-time deployment, that the brigade commander be the reconnaissance team leader, that reconnaissance take place before pre-deployment training commences, that the brigade commander assist the unit commander in the mission analysis, in prioritizing training requirements based on that analysis, and in conceptualizing, resourcing, and conducting unit exercises that will confirm that the requirements have been met. p. 66

xix Recommendations

Battalions, Regiments, Air Squadrons, and Other Similar Size Units

RECOMMENDATION 34: It is recommended that unit pre-deployment training time period be evaluated to ensure adequate generic peacekeeping training on subjects such as the law of armed conflict, negotiation procedures, low-level conflict resolution, and stress management, as well as mission-specific training on subjects such as concept of operations, rules of engagement, standing operating procedures, knowledge of theatre environment, and cultural awareness. p. 67

RECOMMENDATION 35: It is recommended that units warned for deployment be fully supported by subject matter experts. These experts could come from the Land Force Command centres (e.g., the Peace Support Training Centre), be provided by the areas, and, as a matter of practice, come from units that have recently completed a tour of duty in the same peacekeeping mission. p. 67

RECOMMENDATION 36: It is recommended that much more effort be made by areas, brigades, and units to integrate non-military aspects of the UN mission (e.g., NGOs, UN agencies, CIVPOL) into the pre-deployment training, thereby making the peacekeeping partnership a true partnership. p. 68

Canadian Forces Command and Staff College, and Canadian Land Force Command and Staff College

RECOMMENDATION 37: It is recommended that the staff colleges increase their peacekeeping content by modifying their curriculum to include more non-traditional military training for peacekeeping, and to teach selected other subjects in a peacekeeping context. In addition, the colleges should include training with the other peacekeeping partners (CIVPOL, NGOs, UN agencies). p. 68

Army Lessons Learned Centre and Proposed Peace Support Training Centre

RECOMMENDATION 38: It is recommended that the Army Lessons Learned Centre be provided with sufficient resources to collate and update relevant mission area(s) information and intelligence, and provide this in a user-friendly format to individuals and units for pre-mission training. In addition, the lessons learned analysis output of the centre should be regularly transmitted to all training centres for inclusion into or correction of existing training. p. 69

RECOMMENDATION 39: It is recommended that the proposed Peace Support Training Centre be created as soon as possible, and that it be provided with sufficient resources to collect, create, or identify where to find a broad range of training modules, resources, subject matter experts, etc., and that all these be offered to individuals and units for premission training. An ancillary role would be to offer these same resources to other training establishments. p. 69

RECOMMENDATION 40: It is recommended that a review be undertaken to determine the feasibility of amalgamating the responsibilities of the JTF Headquarters, the Army Lessons Learned Centre, the Peace Support Training Centre, the Army Simulation Centre, and perhaps the Canadian Land Force Command and Staff College, under one commander who would have accountability to Land Force Command for army matters, and to NDHQ for tri-service matters. p. 70

Royal Military College

RECOMMENDATION 41: It is recommended that Royal Military College create peacekeeping academic credit courses, and that it seriously consider creating an undergraduate degree in peacekeeping studies. p. 70

RECOMMENDATION 42: It is recommended that RMC cadets receive a minimum of two hours per year solely on the law of armed conflict. This, along with additional training in ethics and dealing with prejudice, should focus on their operational applications, rather than legal or theoretical overviews. p. 71

Canadian Forces Recruit School and Other Basic
Training Establishments

RECOMMENDATION 43: It is recommended that all basic training establishments enhance their training on the law of armed conflict. p. 71

xxi   Recommendations

Battle Schools

RECOMMENDATION 44: It is recommended that the area battle schools be formally tasked to conduct peacekeeping training based upon a Land Force Command curriculum. These schools should also be given a mandate to assist other Commands that have peacekeeping tasks (e.g., helicopter squadrons), based on a curriculum developed by those Commands. p. 72

Reserves

RECOMMENDATION 45: It is recommended that the Reserves, in particular the Militia, review their training objectives with a view to including generic peacekeeping training. In addition, peacekeeping training standards need to be developed to support those objectives. p. 72

RECOMMENDATION 46: It is recommended that Land Force areas assume full responsibility for the enhanced training of Militia augmentees, so that those individuals arrive at a deploying unit at the beginning of the pre-deployment period at the same level of general-purpose combat training and generic peacekeeping training as the unit personnel. p. 73

Office of the Judge Advocate General

RECOMMENDATION 47: It is recommended that there be a Chief of the Defence Staff directive to set out CF doctrine on the law of armed conflict, to emphasize the importance of training in the law of armed conflict, and to identify the Office of the Judge Advocate General as the responsibility centre for training on the law of armed conflict. p. 73

RECOMMENDATION 48: It is recommended that all existing and future law of armed conflict training be primarily focused on integrating it into an operational context, and that operational military such as infantry officers and senior NCMs be trained to deliver much of that training. p. 74

RECOMMENDATION 49: It is recommended that the behavioural aspect of the law of armed conflict be recognized so as to make its teaching an integral part of basic training for all CF personnel, and that there be regular refresher training on this "basic attitudinal training" on the law of armed conflict. p. 74

RECOMMENDATION 50: It is recommended that the Office of the Judge Advocate General be tasked to identify the type and level of special training required for those exercising command functions that are reasonably likely to involve them in dealing with the interpretation and application of the law of armed conflict. In peacekeeping, such individuals invariably include corporals and sergeants. p. 74

RECOMMENDATION 51: It is recommended that the Office of the Judge Advocate General be tasked with overseeing the creation of mission-specific law of armed conflict training that would consist of short refresher courses with particular adaptations or guidance on its application for a particular peacekeeping mission. p. 74

RECOMMENDATION 52: It is recommended that the Office of the Judge Advocate General, in conjunction with various CF training establishments, update or create training curriculum and resources. The Office of the Judge Advocate General should also be encouraged to complete the rewrite and publishing of its 1986 draft Law of Armed Conflict Manual. p. 75

RECOMMENDATION 53: It is recommended that the Office of the Judge Advocate General draw upon the expertise and involvement of the International Committee of the Red Cross in the design and delivery of law of armed conflict training. p. 75

OTHER SOURCES OF TRAINING PRACTICE AND GUIDANCE

Pearson Peacekeeping Centre

RECOMMENDATION 54: It is recommended that the Canadian Forces continue to send its members to all of the Pearson Peacekeeping Centre's courses in order to: train CF members; gain additional expertise to develop CF peacekeeping training; train in the centre's unique civilian-military training environment; and contribute to the civilian-military character and content of the centre's training. p. 76

Canadian Civilian Police

RECOMMENDATION 55: It is recommended that the Canadian Forces look at the training provided to Canadian police in the determination of the minimum use of force necessary, the broad range of use of force options, and how to gradually escalate and de-escalate through this range. p. 78

RECOMMENDATION 56: It is recommended that the Canadian Forces look at the potential usefulness of some Canadian police training in areas such as mediation and negotiation, and officer survival training, including pre-deployment mission-specific officer survival training provided to Canadian CIVPOL. p. 78

Auditor General

RECOMMENDATION 57: It is recommended that the Canadian Forces monitor the Auditor General's current sub-audit on training for peacekeeping for lessons and ideas. p. 79

Other Countries' Military Training for Peacekeeping

RECOMMENDATION 58: It is recommended that Canada follow closely the UN Department of Peacekeeping Operations study on peacekeeping training by all member states, and that the Canadian Forces use that and other indicators as guidance on other militaries' peacekeeping training, in order to assess them in depth and with a view to adapting those programs and materials to Canadian peacekeeping training needs. p. 80

United Nations

RECOMMENDATION 59: It is recommended that as the Canadian Forces develop new non-traditional military training for peacekeeping, they share course packages, training materials, etc. with the UN and other troop contributing nations. p. 82

RECOMMENDATION 60: It is recommended that Canada continue to play a major role in assisting the UN Department of Peacekeeping Operations through such vehicles as training assistance teams, to develop and deliver training standards, materials, and assistance to a variety of other troop contributing nations. p. 82

CHAPTER ONE

Introduction: Rationale and Policy

THE RATIONALE FOR NON-TRADITIONAL MILITARY TRAINING
FOR PEACEKEEPING

Canadian peacekeepers have been particularly adept at peacekeeping. This comes as no surprise in light of the social and educational background of individuals enlisting in Canada's armed forces, combined with the professionalism and general combat readiness of the Canadian Forces (CF). Thus Canada has had a pivotal role in the creation and evolution of peacekeeping. Paradoxically, there has been a belief by many senior Canadian military that traditional general-purpose combat training for war fighting remains sufficient for peacekeeping. This philosophy has prevailed until recently and, as of September 1995, there was little evidence of what this study refers to as non-traditional military training for peacekeeping (elaborated upon below).

However, the study team was made aware of repeated requests by Canadian military personnel for non-traditional military training for peacekeeping, particularly by senior personnel in units about to deploy. Many of those about to deploy or recently returned from peacekeeping missions expressed concerns[1] about insufficient peacekeeping-specific training for themselves and those under their command. Many officers and senior non-commissioned members (NCMs) who have served on peacekeeping missions have concluded, in light of their United Nations (UN) experience, that some specific peacekeeping skills beyond traditional combat readiness are needed to enhance success in peacekeeping. In addition, they feel that some of these skills can and should be taught in advance as opposed to during last-minute pre-deployment training or, even more problematic, in-theatre.

These operational perspectives and requests take on an urgency that the study team found to have been largely unaddressed by the CF training establishment. However, it does appear that there are a number of centres of responsibility within the Department of National Defence (DND) that are about to take, or wish to take, some dramatic steps forward. Several training institutions mentioned that course participants had questioned the low priority and cursory coverage of various peacekeeping topics, and that some limited changes had already been made to meet this "consumer demand." What is key to CF training institutions and mechanisms (e.g., unit training) meeting this demand is a degree of delegated authority to increase or insert such peacekeeping training components. This authority must be combined with sufficient resources to implement this more comprehensive non-traditional military training for peacekeepers.

At the same time, a number of DND individuals and responsibility centres remain sceptical of the need to proceed very far beyond general-purpose combat training. It is perhaps useful to look at the rationale for non-traditional military training for peacekeeping, and the evolution of Canadian and UN policy.

Quite apart from evolving Canadian defence priorities, the UN has witnessed a major change from traditional peacekeeping[2] operations, and is now involved in what is often referred to as "second-generation" peacekeeping. This reflects the growing recognition that the emergencies being dealt with by the UN are complex, and require complex approaches if they are to be resolved in whole or in part. It is no longer assumed that a crisis can be resolved merely by throwing in large numbers of disciplined troops well schooled in the skills of war fighting. Somalia perhaps was the most telling example of the failure of the traditional approach to peacekeeping.

Also, the Somalia operation may have resolved an underlying concern of many Canadian military about accepting peacekeeping as one of their primary roles along with attendant peacekeeping-specific doctrine and training. Some Canadian military have not wanted to see their combat readiness eroded by what they thought might become an unbalanced focus on relatively benign Cyprus-style peacekeeping operations. This perhaps explains the oft repeated phrase that *combat training provides personnel with the complete range of skills needed for peacekeeping operations.* However, and paradoxically, not only have recent UN peacekeeping operations highlighted the need for substantial non-traditional military peacekeeping training, but some of them, such as those in the former Yugoslavia and Somalia, have emphasized the absolute need for combat readiness for peacekeeping.

This study takes as a given that general combat readiness remains the fundamental strength of the Canadian military, and that peacekeeping training extrapolates from, and builds upon, such general-purpose combat training. The very growth in number and complexity of UN peacekeeping operations has progressed alongside increased danger for UN personnel, both military and civilian. Calls for greater UN intervention in internal conflicts will only result in UN missions being placed in even greater danger. It is essential that Canadian troops be ably trained to handle combat situations so as to effectively defend themselves and others in the mission area.

As a result, this study strongly maintains that DND is not facing an "either/or" dilemma, and that in fact both facets, combat training and peacekeeping training, are essential. Aside from both being needed, their symbiosis or strengthening of each other is important. What is immediately obvious is how a combat ready capacity makes Canadian troops better placed to carry out many of the non-combat peacekeeping roles. Less immediately obvious but of equal force is the fact that peacekeeping skills enhance the very security and combat capacity of Canadian troops. For example, the ability to mediate or negotiate out of a situation has obvious security implications for Canadian military, both individuals and units. Similarly, language training and cultural/political knowledge have intelligence gathering implications that can provide tangible tactical advantages. One study, albeit in a limited examination, found that an "analysis of peacekeeping casualties establishes a correlation between specialized training and casualty rates: specialized peacekeeping training seems to reduce casualties among UN forces."[3]

The question remains: What is included in peacekeeping training — i.e., non-traditional military training? The best approach to this is to look at the list of skills required to meet possible operational or functional demands of a UN peacekeeping operation. While not completely unique to peacekeeping operations, these skills are at least substantively different from the skills and attitudes usually taught to Canadian military.

For the purposes of this study, non-traditional military training for peacekeeping refers to the training of skills required in the following situations:

Strategic/Political:

1. Operational limits imposed by UN mandates;
2. Operational uncertainties resulting from imprecise UN mandates and rules of engagement;
3. Working under UN field operational control while remaining under the command of Canadian authorities;

Operational:

4. Different, and at times confusing or minimal, UN standing operating procedures;
5. Working with and alongside non-NATO military contingents and CIVPOL;[4]
6. Working alongside large non-UN international agencies, e.g., NGOs, ICRC;[5]
7. Working alongside major UN agencies, e.g., UNHCR, WFP, UNDP;[6]
8. Working alongside other substantive UN civilian components such as a human rights division or a legal division;

Theatre Environment:

9. Having to deal directly with foreign populations and authorities, particularly at the tactical or community level;
10. Having to deal with armed parties to a conflict as non-enemy albeit at times with a very real possibility that they could become the enemy for reasons of self-protection or UN Charter Chapter VII art. 42 use of force;

Specialized Training:

11. Conflict mediation and resolution;
12. Dealing with issues of human rights violations, e.g., monitoring human rights violations;
13. Dealing with issues of humanitarian assistance;
14. Playing a role in post-conflict rehabilitation;
15. Critical incident stress management; and
16. Operational knowledge of the law of armed conflict.

Dealing with such situations and complications calls for training for UN peacekeeping in addition to general combat readiness training. These skills, and acquiring them, will be addressed more fully in Chapter 3.

Canada has not been alone in its past reluctance to train specifically for peacekeeping, as "...Western military establishments are currently not prepared or trained to deal effectively with the diverse cultural and economic challenges associated with humanitarian operations in developing countries. ...They should retain and promote officers whose expertise includes peacekeeping, humanitarian administration and civilian support operations, an area which is not a career 'fast track' in most military organizations."[7] Some countries, such as those in Scandinavia, have been more enthusiastic in embracing peacekeeping-specific training. The study team noted that many of the CF personnel interviewed were anxious to have the authority to emulate and even move beyond such peacekeeping-specific training initiatives.

One area of training usually seen as traditional military training, and of particular importance in light of the conduct of various UN peacekeeping national contingents in Somalia, is the need for training in the law of armed conflict. "[It] must no longer be regarded as a marginal matter but must be integrated into everyday military life. Respect for the law of armed conflict is a matter of order and discipline. It is the responsibility of leaders to give effect to it and to take it into account in the missions assigned to their subordinates...".[8]

Frits Kalshoven, writing about militaries in general, puts it well in that it "would be a sheer miracle if all members of the armed forces were angels, or even simply law-abiding combatants — and if they remained so through every phase of the war. Factors such as *insufficient or wrongly oriented training programmes* or a lack of discipline may play a role in this respect."[9] (emphasis added)

Canada has been blessed with peace, and the Canadian Forces have inevitably given little training priority to something they began to see as having no short- or medium-range operational applicability. The further historical distancing of World War II and the Korean War, the less relevance the law of armed conflict appeared to have to the day-to-day operations of the Canadian Forces. Simply put, DND largely forgot to take the necessary steps to ensure that their personnel were sufficiently educated in these laws. This oversight took on more serious implications in the light of two related developments. One was the evolution in the content, interpretation, and application of the law of armed conflict. The other, and more important change, was the emergence of a more complicated set of variables faced by peacekeepers in second-generation peacekeeping missions dealing with complex emergencies.

For these reasons, training in the law of armed conflict has been included in this analysis, although unquestionably it should also be a component of traditional military training.

CANADIAN GOVERNMENT POLICY

Parliament

Reflecting the changing post-cold war security requirements and roles for Canada, a number of recent parliamentary committees have looked at peacekeeping: the Senate Subcommittee, reporting in February 1993;[10] the House of Commons Standing Committee in June 1993;[11] and a Special Joint Senate/House of Commons Committee in October 1994.[12]

Certainly the first two committees identified and disagreed with the obvious reluctance of the Canadian military to give added priority to peacekeeping. The Senate Subcommittee quickly established that "a consistent theme in Canadian defence policy has been that participation in peacekeeping operations is not a primary role of the Canadian Forces but a derived task."[13] The report went on to state that:

An internal DND study on peacekeeping recently concluded that 'the tools of soldiering are the tools of peacekeeping ... training them for war trains them for peace.' Thus the best peacekeeper is a well-trained soldier, sailor, or airman, one who knows his or her trade. However, while the Subcommittee accepts that the military training Canadian peacekeepers receive is of exceptional quality, it is persuaded that it could be improved by adding to the curriculum certain subjects which are not necessarily military in character. The entire thrust of this report is that the world has entered a new era, and peacekeeping a new paradigm.[14]

While the Senate Subcommittee agreed as to the importance of general-purpose combat capabilities, it recommended that "more emphasis be placed on dispute settlement and conflict management programs, as well as on the United Nations, regional organizations and [on] peacekeeping history and practice...[CF personnel] should be specifically instructed in the history, tradition, and culture of the country to which they are being sent."[15]

Reporting just four months later, the House of Commons Standing Committee repeated many of the same themes, *inter alia* that

the preparation given to military personnel prior to their deployment in a UN operation:
a) Be provided on a more systematic basis;
b) Be improved to make personnel more sensitive to the different cultures, customs and practices of local populations;

c) Be improved to ensure that all military personnel in units which may be deployed in UN operations receive better training in conflict resolution, mediation and negotiation.[16]

The Special Joint Committee report, the following year, did not focus on training needs *per se*, but did call for "forces that are *flexible, mobile and adaptable*: ... trained and equipped to operate, as recent experience has shown, in circumstances that we would have not have predicted ten or even five years ago — in conditions of near-conflict; in tropical environments as well as northern ones...".[17]

Department of National Defence

The most recent official policy position taken by the Government of Canada and the Department of National Defence was the 1994 Defence White Paper. Only three paragraphs in the White Paper dealt with military training *per se*, stating:

The Government believes that *combat training* ... remains the best foundation for the participation of the Canadian Forces in multilateral missions [peacekeeping]. ... such training equips Canadian Forces personnel with *the complete range of skills that may be needed* to meet the varied demands of the unexpected situations they will encounter.[18] (emphasis added)

This was immediately qualified with a recognition of the "value of cultural sensitivity, international humanitarian law, and dispute resolution training," and then, in what appears to be something of an overstatement, the declaration: "Such training has always formed part of the preparation for Canadian peacekeepers sent abroad; it will be further enhanced."[19]

While such training has not always been provided for Canadian peacekeepers, the future looks promising. The study team found, among those interviewed, that there was a general recognition of the major training gaps in peacekeeping training, and that those involved in training were committed to, and often excited about, addressing non-traditional military training needs.

Aside from stated ministerial positions on peacekeeping training, a number of internal DND studies and an extremely important December 1993 Deputy Chief of the Defence Staff directive exist. These studies and directive are discussed at length in Chapter 2, inasmuch as they pertain more to the internal evolution of Canadian Forces doctrine and operating procedures than to ministerial policy.

It is also worth noting that Canada (Foreign Affairs and International Trade (DFAIT) and National Defence) has just completed a year-long joint study[20] on improving UN peacekeeping rapid reaction. The essence of this study is that the UN, and, by extension its member states and their standby military units, must be much better prepared to deploy rapidly. The traditional situation where Canada and other troop contributing nations had substantial lead time to identify troops and prepare them for deployment will continue to diminish. Rapid reaction will be of increasing importance for UN peacekeepers, and requires advance training as a fundamental.

UNITED NATIONS

In recognition of the lack of UN resources and the scope of training needed for peacekeeping forces, almost all UN statements on training begin by stating that "Member States are responsible for training their national civilian, police and military personnel for participation in peacekeeping missions."[21] At the same time, the UN is very much aware that "increasingly, however, Member States offer troops without the necessary equipment and training."[22] To partially address this training need, peacekeeping training activities are carried out by a number of UN agencies, in particular the Department of Peacekeeping Operations' (DPKO) Training Unit. Others include the training unit of the Office of Human Resources and Management, the Department of Administration and Management, the Department of Humanitarian Affairs, the Training Centre of the International Labour Organization, the United Nations Institute for Training and Research (UNITAR), and the Office of the United Nations High Commissioner for Refugees (UNHCR).

A wealth of expertise exists within DPKO and other parts of the UN. However, invariably they have limited resources, and due to the logistical and political difficulties of producing training materials within the UN, most of the organization's publications and training modules are less advanced than Canadian publications, or are not totally appropriate. *Inter alia*, there would be an advantage in Canada improving its peacekeeping training materials and training programs, and then offering them to the UN as possible models for the training of other national forces.

Basically, the UN expects Canada to design and deliver peacekeeping training for its own military, and is not in a position to provide much guidance or advice on doing so. Indeed, Canada should consider itself as a training resource for the UN.

IMMINENT CHANGE

Until the stresses and strains of peacekeeping missions such as Somalia or the former Yugoslavia, the partial inadequacy of general-purpose combat training for peacekeeping was not very apparent to outside observers or to those back at headquarters in Canada. Obviously, Canadian military personnel who have served in such new second-generation peacekeeping missions dealing with increasingly complex emergencies and often "failed states" have been among the first to identify training shortcomings. Many of them have been calling for added non-traditional military training for Canadian peacekeepers.

Understandably, as Chapter 2 sets out, the Canadian Forces system as a whole has been slower to recognize and adapt to its changing UN peacekeeping role and its attendant peacekeeping training needs. Despite this fact, the study team was encouraged to find that some key components of the CF training establishment have recognized the need for change, and are impatient for the authorization to instigate such change.

In what is hoped will be seen as a contribution to the Commission of Inquiry, and to the many Canadian military who wish to see Canadian peacekeeping training evolve and expand to more effectively meet the needs of modern peacekeepers, Chapter 3 makes observations and recommendations on how existing CF training can be modified or supplemented.

CHAPTER TWO

Training in the Canadian Military

CANADIAN FORCES TRAINING FRAMEWORK

The Defence White Paper provides the Department of National Defence with the governmental guidelines within which policies and programs are developed. Manpower acquisition and its management process are part of that activity. "The CF Personnel Management System ... is a 'cradle-to-grave' process commencing with the identification of a manpower need by a user; progressing through recruitment, training and development of a member on a career basis; and terminating with the member's release. Its primary aim is to ensure that identified manpower needs for operations are satisfied. A secondary aim is the management of careers."[1]

Training is an integral part of that Canadian Forces personnel management system. All Canadian Forces training is aimed at achieving operational readiness, leading to the accomplishment of mission tasks. There are two categories of training: individual and collective.

Individual training is the process by which individual members of the CF acquire and maintain the required attitudes, skills, and knowledge to perform effectively throughout their career, progress from one occupation level to the next, and meet occupation specialty qualifications. Collective training is aimed at developing, maintaining, or improving operational readiness, and generally involves the interaction of two or more units. The two types of training are managed separately, with responsibility divided between two NDHQ groups: Assistant Deputy Minister Personnel (ADM(Per)) and Deputy Chief of the Defence Staff (DCDS). Individual training comes under the purview of ADM(Per), whereas collective training falls within the terms of reference of the DCDS. The planning and management of collective training are further delegated to the functional commands.

Individual Training

CF individual training is primarily aimed at preparing members to meet the job requirements as defined in occupational specifications. To achieve this it is necessary to take into account the broader continuing needs of the organization as well as the career development of the individual member. Expressed in other terms, the individual training process may include elements of:

a. Training — to initiate or improve the desired job performance.
b. Development — to prepare members to assume increasing responsibility within the organization.
c. Education — to build on the members' potential to respond to future needs and goals.[2]

This cradle-to-grave process is designed to take unskilled recruits and provide them with the general military training they will need as servicemen or servicewomen (a continuous process as they progress in rank), and to develop the occupational skills they will require in their trade (also a continuous process as they progress in their career occupations).

There are three types of individual training:

a. General military training;
b. Military occupation training, which includes both basic and advanced training; and
c. Specialty training.

General military training is based upon the General Specification Officers and the General Specification Non-Commissioned Members. It is required for all personnel at specified ranks. For example, junior leader training is required by corporals as a prerequisite for promotion to master corporal. Military occupation training is based upon the specifications for that occupation. Normally, for NCMs, the advanced military occupation training is linked to rank. For example, Qualification Level 6A training is given to master corporals to establish eligibility for promotion to sergeant in a specific occupation. Specialty training is based upon specifications which apply to less than 20 per cent of a military occupation's total establishment. Only those individuals who require the training to do their jobs will receive the training.

There is a close inter-relationship between this "general" training and "occupational" training as personnel progress through their careers.

Individual training is the responsibility of the ADM(Per). ADM(Per) policy directs that "All individual training shall be planned and conducted in accordance with the CF Individual Training System (CFITS), a management model developed to control the quality and quantity of training and to identify the resources for its conduct."[3] The CFITS employs a five-phase approach to training: analysis, design, conduct, evaluation, and validation. In determining what is taught to CF personnel, the analysis phase is the most important: its end product is a course training standard (standards to be achieved at a school or training establishment) or an on-job training standard (standards to be achieved at a workplace). These standards control the quality of training by defining the general aim of the training relative to CF operational needs, and describing the minimum acceptable standards of performance expected.

Basic Training. Basic training is the initial training received by new-entry recruits, either privates or officer cadets. It is conducted at the Canadian Forces Recruit School, the Canadian Forces Officer Candidate School, and the Royal Military College. In the case of the Land Force, some additional basic training is conducted at battle schools. Basic training consists of general military training, such as weapons, drill, basic field craft, etc. It also begins the inculcation of personnel into the military, by teaching the basic components that make up the military ethos (for example, principles of leadership, personnel policies, the law of armed conflict, military law, etc.).

Advanced Individual Training. Advanced individual training may be conducted throughout a military career, and at many different locations. As noted earlier, it can be general military training, military occupation training (frequently a mixture of the two), or specialist training. For example, leadership and staff courses will be conducted for sequential rank levels (regardless of occupation) that are general military training in nature, but at ever-increasing advanced levels. Military occupation and specialist training are generally occupation-specific, and are also conducted at sequential rank levels.

Refresher Training. This individual training is generally conducted annually and is usually Command-level directed. In the case of the Land Force, it is regulated by individual battle task standards. The standards require that all tasks are tested prior to refresher training; if a deficiency is identified, the individual would receive refresher training.

Individual Training Management. Responsibility for all aspects of the individual training system is divided between three levels of management:

a. NDHQ — establishes policy, determines quantitative needs, describes occupation specifications and verifies policy implementation;
b. Designated command/training agency — produces training standards, validates training and provides resources; and
c. Training establishment — designs and conducts training and evaluates course members and training.

Collective Training

Collective training focuses on team or unit performance to achieve and improve operational readiness. Its policies are the responsibility of the Deputy Chief of the Defence Staff, whereas the planning, development, and management of collective training are delegated to functional commands through the Chief of the Defence Staff (CDS) Direction to Commanders, which details the missions to be accomplished by the CF, including training missions.

Collective training takes place at the level of units and formations. It is the bringing together of the basic skills of the individual soldier so that the individuals operate as a team (sub-unit) and the sub-units operate as a cohesive unit. Practised collectively as a unit or number of units, it incorporates both general military training and occupational training and can be oriented to almost any generic or mission-specific scenario: traditional war fighting, generic peacekeeping, a specific type of peacekeeping mission, aid to the civil power, etc.

It is during mission-specific collective training that the vast majority of the Canadian Forces peacekeeping training takes place.

Canadian Forces Training Principles

CF personnel receive their career development through a combination of formal training and education, and on-job experience. This begins with basic training on general military training subjects (including an introduction to the military ethos), followed quickly by occupational training. As individuals demonstrate capability and potential on the job, they receive further training to prepare them for additional leadership responsibilities. The additional rank invariably means additional occupational knowledge requirements. This close relationship between formal training and the development of individuals is prevalent throughout their careers to the point that neither training nor career development can be viewed separately: the system is an integrated one.

Training within the Canadian Forces is a matrix of training at the individual level (including basic training) and collective level, conducted either as a specific event or as part of refresher training. Within this matrix, the training may be either general military training or occupational training. Inherent in the concept is "that training tasks must match job tasks as closely as possible."[4]

Since peacekeeping training can occur in any part of this matrix, the study team identified key or representative units or establishments that are or should be part of CF peacekeeping training, be it delivering training or authorizing it. *Inter alia* this included:

- National Defence Headquarters;
- Land Force Command Headquarters;
- a Land Force area headquarters;
- a brigade group headquarters;
- a number of battalions and sub-units;
- Canadian Forces Command and Staff College;
- Canadian Land Force Command and Staff College;
- Army Lessons Learned Centre;
- Royal Military College;
- Canadian Forces Recruit School; and
- Office of the Judge Advocate General.

It was believed that this mixture would provide a representative overview of peacekeeping training in the Canadian Armed Forces.

BACKGROUND FOR CANADIAN FORCES PEACEKEEPING TRAINING

Before the end of the cold war, the Department of National Defence conducted several policy reviews to reconcile doctrine and training with changes in the national and international fiscal and security environments. A major evaluation of Canada's contribution to peacekeeping was conducted in 1981, and the recommendations of Evaluation E1/81 still form the basis of much of the policy in effect in DND today.

In addition to the 1981 evaluation, a number of staff reviews and studies of peacekeeping within DND took place. Two principal studies which clearly reflect the opposing concepts for preparing for peacekeeping are:

a. FMC 3450 - 1(COS OPS) 16 May 1984 - FMC Paper on Peacekeeping (often referred to as the Lalonde Study after its principal author, BGen. P. Lalonde, then Chief of Staff of Mobile Command). This study concluded that military forces organized, equipped and trained for combat (i.e., a general-purpose combat capability) are the most effective for the peacekeeping role.
b. DCDS Appreciation of the Situation - CF Peacekeeping Resources and Commitments 19 Sep 1989 (often referred to as the Rowbottom Study after the NDHQ staff officer who headed the study). This study concluded that the CF ought to create specialized forces and procedures for the peacekeeping role.

The debate over the two concepts — general-purpose or specialized — has political and military implications, and has continued in many forums to this day. It is an important and evolving topic. Both Chapter 1 and Chapter 3 of this study examine several aspects of the debate, ranging from dedicated training facilities, to upgrading current CF courses, to balancing general training with mission-specific training.

NDHQ Program Evaluation E2/90

In November 1990, the DND Deputy Minister and the Chief of the Defence Staff directed that another departmental evaluation take place; it was not completed until June 1992. Known as NDHQ Program Evaluation E2/90, the aim of the evaluation was:

a. To review the relevance and effectiveness of DND's peacekeeping operations;
b. To examine specific aspects of program design, delivery, and control; and
c. To study alternatives that could enhance, in a cost-effective way, the achievement of Canada's foreign and defence policy.

A number of the evaluation's conclusions have training implications, and are therefore pertinent to this study. These include:

a. Standardized command and control and communication systems across the Canadian Forces for peacekeeping do not exist. For each operation, Canada initially uses *ad hoc* arrangements and thereafter puts in place an operation-unique system.[5]

17 Training in the Canadian Military

b. The Department's central training organization, Canadian Forces Training System (CFTS), does not play a major, central, or coordinating role in the provision of peacekeeping training and education. While there has been no detailed, in-depth analysis on the establishment of a Canadian centre for peacekeeping training, there was also no support for it at this stage. *However, given the interdependence of potential peacekeeping missions, and national and international economic, political and security issues, peacekeepers will need more than only conventional General Military training.*[6] (emphasis added)

c. The Evaluation Team found next to no support for the option of establishing specialized peacekeeping forces among those interviewed, military and non-military, at home and abroad. Indeed the general consensus was that peacekeeping is and should remain just one more activity that general purpose forces must be prepared to carry out on a 'contingency basis'.[7]

d. In view of the current international conditions, the need for multiple tasked, general purpose military forces is reinforced. In terms of contingency force operations, it is feasible to scale-down a force to meet peacekeeping requirements, but impractical to do the reverse.[8]

e. It is important to differentiate between general military training and specialized peacekeeping training. All sources agreed that the CF must give priority to maintaining a general purpose military capability. Moreover, general military skills are also essential to execute peacekeeping missions successfully. *Therefore, peacekeeping training should be regarded as an overlay which concentrates on the UN (or other sponsoring agency) 'system,' the force mandate, geopolitical/intelligence briefs and special techniques such as those required for observers.*[9] (emphasis added)

f. ... the new peacekeeping environment would be well served by selected improvements in training. These could include: a general knowledge package imparted to recruits and officer cadets on entry; coordination of material taught at all levels of staff colleges and CF schools to include focusing on typical problems encountered in all phases of peacekeeping operations; and the improvement of guidelines and briefing programs for Canadian Contingent Commanders and officers appointed to the staff of a peacekeeping force.[10]

g. Shortfalls in training exist at UN Headquarters and national levels.[11]

DCDS Directive 4500-1 (DCDS) 29 December 1993

As noted in Chapter 1, both the House and the Senate conducted reviews on peacekeeping in 1993. Both had specific recommendations on training, some of which have been set out in Chapter 1. At about the same time, J3 Peacekeeping at NDHQ conducted a survey[12] of Commands, colleges, and schools to assess how existing training programs and courses develop peacekeeping-related skills compared to the guidelines issued by the UN. In placing the survey in context, J3 Peacekeeping stated that while most of the UN requirements were covered during general military or pre-deployment training, there was a growing concern that not all of them were adequately covered.

These UN guidelines[13] included:

a. Geopolitical studies (geography, history, economy, political system, defence and internal security forces, internal and external influences);
b. UN mandate and mission (dispute history, major developments, civilian involvement);
c. Administration, medical, pay, etc. (pre-deployment preparations);
d. Weapons training (personal, crew-served, others);
e. Equipment recognition (internal and external threats);
f. Navigation equipment, radar;
g. Chemical warfare (characteristics, symptoms, precautions, use of equipment);
h. Mine awareness;
i. Field exercises and battle procedure;
j. General military training (physical training, communications, first aid, hygiene, health awareness);
k. UN operations (observation duties and posts, checkpoints, roadblocks, searches, patrolling, investigations, negotiations, liaison, use of force, leadership);
l. Safety measures; and
m. Special training (driving, armoured-vehicle drills, helicopter emplaning, staff training, explosive disposal, media relations).

The responses to the J3 Peacekeeping survey reflected a cross section of traditional training being conducted but, in general, few of the formations and institutions were conducting specific UN training or education. The study team found the following to be of note:

19   Training in the Canadian Military

a. Maritime and Air Command had little training that correlated to the UN requirements;
b. Land Force Command (LFC) felt that the concern mentioned in the preamble to the J3 questionnaire was "completely unfounded" and that the training being conducted by that Command was meeting the requirement;
c. No Canadian Forces Training System training specific to UN requirements was being conducted, including at the Recruit School;
d. No military courses specifically aimed at meeting the peacekeeping requirements were being conducted at the military colleges. Some relevant military training was being given, and some training was available in the graduate program if selected by the candidates;
e. The Canadian Forces Command and Staff College conducted six lectures, one discussion period, and one exercise on peacekeeping; and
f. The Officer Professional Development Program included about 11 hours of self study on peacekeeping or peacekeeping-related subjects.

The combination of the recommendations of the Standing Senate Committee on Foreign Affairs February 1993 report,[14] the Standing Committee on National Defence and Veterans Affairs June 1993 report,[15] the NDHQ Program Evaluation E2/90 (issued in 1992), and other studies, led the Deputy Chief of the Defence Staff to issue directive 45001 (DCDS) 29 December 1993 (see Appendix I) on training requirements for peacekeeping operations, with a view to rectifying deficiencies, implementing improvements, and formalizing the guidance and direction for preparing and training CF personnel for peacekeeping operations.

In examining the training requirements, the DCDS found:

a. CF peacekeepers get the majority of their information and training from three areas: from general CF courses which cover specific peacekeeping topics; from annual refresher training; and from pre-deployment training. Information and training concerning UN tasks and roles received during CF career courses tends to be general in nature, but essential as a foundation upon which to build.... The principal reason a good soldier makes a good peacekeeper is the core of general military knowledge and skills he brings to annual refresher and pre-deployment training sessions.[16]
b. In that no two missions are likely to be faced with identical mandates and circumstances, the best core training for these diverse

operations has been found to be general purpose military training, where emphasis is on basic skills and specific to classification skills required by officers and noncommissioned members. To augment these skills, there is a requirement, as with any military operation, for periodic refresher training, specialist training that may be dictated by the mission or circumstance, and comprehensive pre-deployment training.[17]

The DCDS came to the following conclusions:

a. It is the accepted dictum that the CF takes trained soldiers, sailors and airmen and provides specialized training prior to deploying them on a peacekeeping mission. This training ... concentrates on theatre operations, the mission mandate and environmental, cultural and administrative preparation for individuals or formed units. It is normally decentralized, directed to the specific mission, and its length will vary depending on the urgency of the deployment and the nature of the task.[18]
b. There are several critical areas which require additional emphasis and reinforcement, especially mission specific ones, just prior to deployment. Also, although extensive experience exists on peacekeeping training in the field, *little formal written guidance exists on course content* other than that contained in UN training manuals and locally produced pamphlets.... Finally, it has become apparent that *the CF has a requirement for a visible focal point for peacekeeping training, standardization and doctrine, a so called 'centre of excellence'* to bring all of this expertise and knowledge together.[19] (emphasis added)

The DCDS instruction states:

a. Since personnel are expected to arrive at the training sessions current in basic military skills, priority is to be given to mission specific information such as the UN policy, ROEs [rules of engagement], the geo-political and military situation and cross-cultural awareness. In the case of formed units, attention must be focused on the development of collective operational skills and cohesion.[20]
b. Additional mission dependent skills, such as mine awareness training, convoy escort duties or cordon and search operations will be taught as required. Verification or instruction of small arms skills,

physical fitness and first aid will also take place during the training programme. The training directive that is issued as a supplement to the NDHQ tasking order will outline any additional requirements.[21]

Examples of the directions the DCDS issued include:

a. Create and establish mechanisms and procedures to ensure an accurate definition of training requirements; effective management practices for the production, distribution, configuration, and control of course-related materials; and a sound instructional analysis capability.
b. Formalize pre-deployment training and packages for individuals, composite units, and formed units.
c. Formalize pre-deployment training so that it concentrates on mission-specific training, an intensive review and confirmation of general military skills and personal and personnel administration.
d. Develop training and general information packages for recruit and basic officer training as well as for junior and senior leaders' courses, the Officer Professional Development Program, staff school, and staff colleges.
e. Implementation of the directive is to be reviewed periodically by J3 Peacekeeping, with an interim report by August 1, 1994, and a comprehensive report by August 1, 1995.

In his instruction, the DCDS included training requirements for, first, military observers, staff officers, and small composite units, and, second, formed units.

The DCDS instruction is an important document that, to the study team's knowledge, remains the latest policy directive. As will be amplified further, despite the very specific tasking and milestones it contains, the study team was unable to find much practical and concrete evidence of its execution,[22] nor any evidence of J3 Peacekeeping reports as described in paragraph e. above.

The failure to respond to the DCDS instruction is as indicative as its policy content, and needs to be examined for lessons on how future changes to peacekeeping training can be implemented. The study team believes that the failure to respond stems both from a misunderstanding of the changes required, and a lack of a sufficiently senior-level CF responsibility centre to provide ongoing guidance for and oversight of such training.

DCDS Study Directive on Peacekeeping Training in the Canadian Forces

DCDS Study Directive (DCDS) 14 September 1995 (see Appendix II) has set in motion a study on peacekeeping training in the Canadian Forces. Its aim is "to evaluate peacekeeping training required in addition to normal combat and occupational training, and to identify the most effective and cost efficient manner of achieving this training." DND has created a working group and it is tasked with "providing overall user guidance and direction leading to the development of a project study plan and for identifying information sources, survey populations, subject matter experts, existing training programmes and related training documents."

The creation of this working group is a welcome initiative, and its potential impact will be substantial if the DCDS lends his continued support to its progress and its recommendations. His personal support is critical inasmuch as the composition of the working group is relatively low level, as is the responsibility centre for the working group, J3 Training. Unfortunately, this situation has been relatively common in the past, with various DND centres of responsibility for peacekeeping having unclear terms of reference and often being situated at a middle management level without the organizational influence to advance initiatives.

Despite this, within NDHQ the general attitude towards the importance of peacekeeping appears to be changing rapidly. Therefore, this DCDS-initiated working group will probably face a more welcoming environment,[23] and its recommendations may have far-reaching impacts.

UN Peacekeeping Training Assistance Teams

Recently Canada has taken a major lead in the creation of training assistance teams (TATs) to assist member states with the establishment or provision of peacekeeping training. This reflects the importance that the UN is increasingly giving to training, and Canada's involvement reflects the importance that the Canadian government gives to the goal of upgrading the training level of un peacekeeping troops. A Canadian colonel has recently taken command of the UN DPKO Training Unit in New York. Canada's lead in this is a logical one in light of the high quality of traditional military training provided to Canadian military.

Paradoxically, however, these TATs will be providing peacekeeping-specific training that goes beyond that currently provided on any regular basis to Canadian peacekeepers. This UN exercise is in the process of identifying the training materials and packages that exist in national militaries, with a view to creating composite packages with the best from each of them. Canada, and in particular the DND working group referred to above, will be well advised to take note of the findings, particularly the materials from various European militaries and New Zealand (NZ mediation/negotiation training).

Doctrine Publications

The study team located two principal doctrinal publications: CFP(J)5(4), Joint Doctrine for Canadian Forces Joint and Combined Operations, produced in June 1994, and B-GL-301003/FP-001, Operations Land and Tactical Air Volume 3: Peacekeeping Operations, produced in November 1994.[24]

CFP(J)5(4) devotes one of 25 chapters to peace support operations (used in a generic sense) to describe activities in international crises, and conflict resolution and management. Paragraph 2111 defines

a set of decision support criteria for [Canada's] involvement in international peacekeeping operations or observer missions. All criteria are not necessarily applied to all peacekeeping operations; rather, they serve as evaluation guidelines for political and military decision makers responding to requests for Canadian peacekeepers. Depending on the type of mission, the following criteria, which are not all-inclusive, may apply:

a. threat to international security;
b. political settlement;*
c. clear and enforceable mandate;*
d. support of a peacekeeping deployment;
e. agreement to Canadian participation;*
f. failure of peaceful means to resolve the dispute;
g. command and control and concept of operations;
h. force's international composition,* size and equipment;
i. funding arrangement and logistics concept;*
j. single identifiable authority;*
k. clear rules of engagement;
l. acceptable level of risk to CF personnel; and
m. interference with other missions.*[25]

* Criteria specified in 1987 Defence White Paper.

The paragraph dealing with peacekeeping training states:

General purpose combat training provides CF personnel with most of the prerequisites for peacekeeping duties. To complement general purpose training programmes, the CF provides specialized training and indoctrination before committing personnel to a peacekeeping mission. This training, which concentrates on theatre operations, the mission's mandate, and environmental, cultural and administrative preparation for individuals and units, has been developed and conducted in accordance with UN training guidelines and includes the following:

a. Unit training: training generally conducted by all units warned of possible participation in a peacekeeping operation. This training normally lasts between 10 and 30 days for composite or formed units and may be extended as required.
b. Contingency training: annual training conducted by the CF UN standby force.
c. Staff Officer and Military Observer training: training conducted over a 5 to 7 day period as required for officers warned of appointments as staff officers or to existing or new observer missions.
d. Rotation or replacement duty training: training conducted over a 7 to 14 day period for personnel proceeding on individual rotation or replacement duties.
e. Seminars: two or three day peacekeeping seminars conducted periodically as required.

The paragraph continues:

The general syllabus for Canadian peacekeeping training follows UN guidelines to its member states. In accordance with these guidelines, the type of training described in the following paragraphs is conducted.

1. Geo-political Briefings: these are briefings on the region in which a peacekeeping operation is conducted and are provided through background reading, lectures and locally produced aide memoires. They include:
    a. geography;
    b. history;
    c. economy;
    d. political systems, government;
    e. defence and internal security forces;

f. internal influences, including religion, militias, revolutionary movements, etc.;
g. external influences; and
h. culture and customs.
2. Mandate and mission study: this training covers the history of the dispute, major developments, civilian involvement, civilian and military cooperation and coordination, and the specifics of the mandate.
3. Legal considerations and the use of force: this training includes familiarization with legal considerations, the Laws of Armed Conflict, the Convention on Privileges and Immunities of the UN, rules of engagement and the use of force.
4. Weapons and equipment training: this type of training covers individual and crew served weapons, familiarity with theatre weapons, vehicles and equipment, night vision equipment, NBC warfare, mine warfare awareness and field exercises.
5. General military training: this training involves physical fitness, map reading, communications and first aid.
6. UN operating techniques: this training provides an awareness of duties and responsibilities with regard to such activities as observation posts, check points, road blocks and searches, vehicle and foot patrols, investigations, negotiations and liaison, use of force and leadership.
7. Safety measures and precautions: this training involves safety awareness with respect to shelters, equipment, travelling and movement, and non operational safety to recognize and prevent health problems (stress and depression), traffic accidents, fire, suicide, weapon handling accidents and accidents on leave.
8. Specialized training areas: this training includes driver training, helicopter training, UN staff training and procedures, explosive ordinance disposal, familiarization and training in media relations.
9. Field exercises and battle procedure: this training is conducted as annual training or during the pre-deployment phase. Where possible, battle simulation is incorporated.[26]

The other doctrinal publication, B-GL-301-003/FP-001 Peacekeeping Operations, has one chapter devoted to training. That chapter outlines the Canadian Forces training concept:

In most cases, the general purpose military training conducted by the CF is to prepare individuals for most peacekeeping tasks. It is only in certain cases, for specific tasks such as negotiation techniques, that additional training has to be conducted. Prior to deployment, the UN Secretariat issues training instructions for the operation. A general training instruction may be issued by LFC HQ or Area HQ. The unit commander and staff review these instructions and establish the priorities for training to be conducted prior to deployment and utilize all available opportunities to conduct the training to satisfy those priorities. The nature of predeployment training is largely determined by the time and training resources available, the peacekeeping experience of the deploying personnel, the size and composition of the force, and the type of peacekeeping operation to be conducted.[27]

The degree and intensity of training a soldier requires for peacekeeping duties is determined by ... the time available, the resources available, and the focus of the commander, based on the tasks to be performed and the skill level of the soldiers selected.[28]

While the exact composition of the training programme varies from operation to operation, the emphasis of all training must be on the following three areas:

a. honing critical skills;
b. conducting unique-to-mission training to ensure the soldier understands what role peacekeepers are to play for that specific mission; and
c. ensuring the attitude and social skills required on the mission are enhanced.[29] (This is later explained to be vigilance, alertness, inquisitiveness, patience, restraint, endurance, initiative, and leadership skills.)

This publication also contains guidelines for preparing and training military personnel for assignment with the UN or other operations. These guidelines include "normal" training (such as physical fitness, communications, weapons, map using, first aid, etc.), "special" training (such as mine awareness, explosive disposal, helicopter, and public information), and "mission-specific" training. This mission-specific training checklist[30] includes:

a. Environmental information (climate, terrain);
b. Historical background (historical/economical/social overview) of the conflict, internal and external influences, background to peacekeeping operations in general (the UN Charter, UN organization, history of peacekeeping), and the development of the mandate;

c. Cultural information (customs and traditions, social values and norms, acts of social custom and politeness, elements of indigenous social psychology, do's and don'ts, basic words and phrases, briefings by area specialists);
d. Military/political background (personalities and leadership, force structures and orders of battle, weapons/equipment, doctrine and training, threat to the peacekeeper);
e. Operations (mission and tasks, force organization, standing operating procedures, security arrangements, mine awareness, detainee handling procedures, negotiations, investigations and liaison);
f. Use of force guidelines (concept, rules of engagement); and
g. Legal information (law of armed conflict, theatre legal framework, status of forces agreement, legal liabilities).

These two recent doctrinal publications, Joint Doctrine for Canadian Forces Joint and Combined Operations June 1994, and Operations Land and Tactical Air Volume 3: Peacekeeping Operations November 1994, are not well-known in the field. Few of those interviewed by the study team were familiar with their contents, and the doctrine has only partially been translated into training directives.

INTERVIEWS

The study team conducted formal visits and interviews[31] with the following Canadian military:

a. National Defence Headquarters
   J3 Peacekeeping Training: LCol. W. Reid
   J3 Operations: Col. C. Ross
   Director Professional Development: Col. P. Holt
   J4 Director General Material Comptollership and Business Management: LCol. D. Lynch
b. Land Force Command Headquarters
   G3 Operations and Plans: LCol. R. Davis and team
c. Land Force Central Area Headquarters
   Commander: MGen. B. Stevenson and team
d. Headquarters 2 Canadian Brigade Group
   Commander: BGen. B. Jefferies and team

e. 1st Battalion The Royal Canadian Regiment
    Operations Officer: Maj. J. Vance and team

f. The Royal Canadian Dragoons
    Deputy Commanding Officer: Maj. G. Hall and team

g. Canadian Forces Command and Staff College
    Director of Joint Studies: Col. D. Taylor and team

h. Canadian Land Force Command and Staff College
    Commandant: Col. S. Labbé and team

i. Army Lessons Learned Centre
    Commander: LCol. P. Cunningham and team

j. Royal Military College
    Commandant: BGen. C. Emond and team

k. Canadian Forces Recruit School
    Commanding Officer: LCol. Lehoux

l. The Royal Canadian Regiment Battle School
    Deputy Commanding Officer: Maj. W. Beaudoin

m. Office of the Judge Advocate General
    LCol. K. Carter, LCol. J. Holland

n. 427 Squadron Petawawa, Op Pivot rotation to Haiti
    Deputy Commanding Officer: Maj. B. McQuaid

In addition to these meetings, team members had informal contact with other Canadian military, and met with a broad variety of UN and NGO members, as well as other interested individuals. In particular, the team met with individuals in the Training Unit and other divisions throughout the UN Department of Peacekeeping Operations, and related parts of the UN Secretariat in New York; the Pearson Peacekeeping Centre; the Royal Canadian Mounted Police; the Canadian International Development Agency; the Office of the Auditor General, and the Department of Foreign Affairs and International Trade.

National Defence Headquarters

For peacekeeping training, National Defence Headquarters is responsible for interfacing with the political authorities, setting departmental and Canadian Forces policy, issuing directives and guidelines to the Commands and units under its command, and providing resources to implement those directives.

It is the view of the Canadian Forces that troops assigned to peacekeeping duties need to be well-trained in conventional military skills and operations, and that the best core training for such duties is general-purpose military training, with emphasis on basic combat and specific-to-occupation skills. This, including annual refresher training in these areas, fulfils the majority of the CF training requirements for unit and individual participation in UN peacekeeping operations. The CF has stated that they recognize there is an additional requirement for a training overlay in UN and mission-specific subjects.[32]

Pre-deployment training of two to four months' duration (depending on the amount of warning time) for formed units, support companies and specialist companies, is organized and directed by the applicable Command headquarters, and conducted at unit or composite unit level. This includes UN and mission-specific subjects. For the training of individuals, such as those required for military observer and staff officer positions, including UN standby personnel, NDHQ organizes, coordinates, and conducts one or two annual training sessions of eight days' duration. These include the UN and mission-specific subjects, and refresher training in vehicle driving, weapons handling, and mine awareness.[33]

However, there does not appear to be a traditional, structured sequence of concept/doctrine/standards/training in existence for peacekeeping. Traditionally, the concept would be a derivative of government and Canadian Forces policy, which would be clearly stated in a joint, national-level approved document. Doctrine to support the concept would be developed initially on a joint basis (perhaps with one environment leading), and then, if necessary, in more detail on a single environment basis. Training objectives and standards would then be set by the responsible levels of command, first to provide guidance on the level of intensity of the training in a degree of priority, and second to evaluate the effectiveness of that training. Flowing from the standards would be the production of training curriculum, training packages, standing operating procedures, *aide memoires*, etc. This traditional procedure is not institutionalized for peacekeeping at the national level.

The study team was unable to identify a central, joint focal point for peacekeeping training at NDHQ. Responsibilities seem to be blurred and are still evolving. For example, some responsibility rests with J3 Training (operations), some with the doctrine writers (force development), and some with the directors responsible for individual training and professional development (personnel). Evaluation E2/90 has many important observations and recommendations, and the DCDS December 1992 instruction gives some direction to implement some of these. However, these are only partially implemented and the study team did not find a focused follow-up to those recommendations. The study team has concluded that in many instances, a gap exists between what NDHQ and the Commands believe is being done, and what is actually being done in the field.

One of the causes for this gap is the lack of a central staff that has access to a "corporate memory bank." An internal NDHQ review conducted in 1994[34] concluded that there were no problems with training for peacekeeping at the unit or sub-unit level (which is coordinated by the Commands), but training assistance for units in UN and mission-specific subjects should be provided by a centralized peacekeeping staff. That review further concluded that the location of this staff should be with other organizations that have related goals such as the Army Lessons Learned Centre (see below) with a view to improving information management and the development of peacekeeping techniques, doctrine, training methodology, and standards.

On May 26, 1995, a proposal was approved by the Vice Chief of the Defence Staff to immediately establish a small peacekeeping training staff of five personnel under the command of Land Force Command. The establishment in Kingston of this Peace Support Training Centre (discussed further below) has been delayed, and it will not be staffed until summer 1996. Its responsibilities will include:

a. Providing instructional assistance to the Commands in the conduct of unit and sub-unit peacekeeping training;
b. Training CF individuals selected for UN military observer staff and other support positions; and
c. Coordinating the development of peacekeeping training and standards, standing operating procedures, training manuals and packages, *aide memoires*, and bulletins for the CF.

At NDHQ, the "battle procedure" (the process of sequencing and allocating time for critical actions) for peacekeeping operations seems to be top-heavy. Too much time is spent by NDHQ documenting the mission, concept of operations, command and control, and rules of engagement into formal documents, at the expense of the units who need this information in order to conduct mission analysis and determine training requirements.

Those institutions and processes directly and indirectly controlled by NDHQ (recruit school, military college, officer professional development) do not appear to be under direction to teach peacekeeping subjects.

Peacekeeping subjects that are beyond the capability of units to resource or teach, and which would seem to be candidates for national level "packages" (e.g., stress management, low-level conflict resolution, negotiation procedures, culture awareness, media relations, etc.), have not been addressed by national-level agencies.

The study team was unable to locate any national-level approved peacekeeping training standards.

Land Force Command Headquarters

With regard to peacekeeping, Land Force Command Headquarters is the principal one responsible for the force generation of army units preparing for deployment on peacekeeping missions. Obviously, it has major training responsibilities, which include implementing NDHQ directives, issuing its own guidance, and providing resources.

LFC has recently issued guidance on UN training. This guidance (LFC 3450-2-4-1 (G3) 18 April 1995) directs training on subjects such as:

a. Theatre familiarization (geography, climate, religion, ethnic diversity, essential/useful phrases, customs, and economics of the region);
b. UN Force organizations;
c. Belligerent force recognition;
d. Rules of engagement (and briefings on the law of armed conflict);
e. Negotiation procedures;
f. Media awareness;
g. Observation techniques; and
h. Internal security.

The LFC Headquarters instruction is good, but perhaps because of its newness, is not fully implemented or understood in the field. Recognizing that each deployment will be different and that the instruction will be adjusted to meet the circumstances, there still remain some uncertainties. There is no guidance on "standards" for UN training or on the balance of time to be spent between general training and mission-specific training. The instruction refers to "packages" (e.g., low-level negotiation, media awareness), which the study team could not locate in the field.

The LFC guidance also directs that all personnel deploying on operations must achieve the Individual Battle Task Standards. These standards are contained in B-GL-304-002/PT-Z04, were implemented beginning 1 April 1995, and cover subjects such as weapons, physical fitness, first aid, field craft, etc. In addition, B-GL-304-002/PT-Z04 makes reference to the production of Battle Task Standards for Operations Other Than War, which would cover United Nations, humanitarian, and internal security operations. It is the study team's understanding that these standards have not yet been developed.

LFC has already created an Army Lessons Learned Centre. It was recently relocated to Kingston, with the intention that it will be co-located with the Peace Support Training Centre due to be staffed by summer 1996. Following its review of LFC's proposed process of connecting policy/lessons learned/doctrine/training, the study team's recommendations are put forward in Chapter 3.

Land Force Central Area Headquarters

This level of Headquarters is primarily responsible for providing the personnel and equipment necessary to implement the mission. This is mainly manifested in the provision of Regular and Reserve Force augmentees to formed or composite units. The area commander is given some responsibility in endorsing the certification of operational readiness of a warned unit.

The area issues its own training guidance to units warned for deployment, which repeats the Command's direction but will also indicate the area commander's emphasis to be placed on certain standards.

Most of the observations of Command level are also applicable to area level. In particular, instructions and resource allocations issued by area Headquarters must be timely.

Headquarters 2 Canadian Brigade Group

This level of headquarters usually has full command of a unit preparing for deployment, and the commander is responsible for certifying the unit as operationally ready. It is the level of headquarters responsible for freeing the warned unit from routine tasking by reassigning that tasking to other units. It is also responsible for providing training resources to the warned unit(s).

It would be rare for any two units to be at the same level of readiness at the same time, and it is therefore rash to believe that "global" training guidance can be applied to most situations. It is the brigade commander who most influences the training of a unit preparing for operations, and does so by:

a. Interpreting the guidance and direction of the national, command and area levels, conducting a mission analysis of the operation in question, and weighing the requirements against a personal knowledge of the readiness of units;
b. Issuing guidance and direction to fit the mission, the concept of operations, and the time and resources available;
c. Providing and scheduling resources for the units;
d. Monitoring pre-deployment training; and
e. Conducting a confirmatory evaluation exercise.

Of particular importance, and interlocking, are the mission analysis and the confirmatory evaluation exercise. Mission analysis is a process of determining the type of forces required, their equipment, and their level of training. Before it can be completed, the brigade commander and the commanding officer need to know:

a The statement of the mission, which leads to the determination of specific and implied tasks;
b. The command and control arrangements;
c The concept of operations; and
d. The rules of engagement.

This information should be forthcoming as soon as the political/military decision to participate in the mission is made, but, unfortunately, this rarely occurs. At the very least, the information needs to be confirmed during the pre-mission reconnaissance. This implies that the reconnaissance must, first, involve the chain of command that determines operational readiness, and, second, be early enough to allow adjustment to the training plan before the training has commenced.

The confirmatory exercise has two purposes: it provides a vehicle for realistic last-minute training and it evaluates the readiness of the unit. To accomplish this, it should involve subject matter experts (preferably personnel who have just returned from the theatre of intended operations or have had similar experience), and training "incidents" that are as close to an actual situation as is possible.

The study team believes that this combination (mission analysis and confirmatory exercise) is probably the most important step in preparing for a specific peacekeeping mission. For it to be complete, it is important that the commander be "up to speed," and for this reason, the commander (or a senior representative) should be part of the reconnaissance process.

1st Battalion The Royal Canadian Regiment and The Royal Canadian Dragoons

The study team visited these two units shortly after they had returned from successful UN rotational tours in Croatia and Bosnia respectively.[35] This was the most useful visit by the study team: it gave the clearest indication of actual training received, as well as the effectiveness, completeness, and appropriateness of that training.

This type of unit would usually receive about six months' warning and be allocated a 90-day pre-deployment training period. As is usually the case, both units needed to be augmented, either to bring their numbers up to the UN requirement, or to provide personnel with skills not usually found in the unit. In most cases, but not all, these augmentees arrived in time to begin the pre-deployment training.

While circumstances for the two units were not identical, comments received from both included:

a. Even though the units were not the initial units and were rotational, their detailed instructions were late in arriving from NDHQ relative to their pre-deployment training. For example, their warning orders stated that training guidance would be issued in the operations orders, but these were not received until after the reconnaissance in the case of one unit and immediately prior to commencement of pre-deployment training in the case of the other. There was no ammunition allocation in the orders, and the units were unclear of their allocation until midway through the preparation phase.
b. In the case of one unit, in-theatre reconnaissance took place about three weeks after pre-deployment training had commenced. This left little opportunity to readjust unit training to fit the theatre situation.
c. Few information "packages" were available. Indeed, the majority of the non-traditional military training was conducted on the initiative of the units themselves on an *ad hoc* basis, using scrounged resources

and subject matter experts found by the units. Instead of having an organized system of modules and packages being automatically "pushed" down to them by responsibility centres, the units had to "pull" for information. In particular, no training packages were provided on the law of armed conflict, negotiation procedures, low-level conflict resolution, and critical incident stress debriefing training. The units did not use any United Nations Headquarters-produced training material. A soldiers' handbook (part general information and part weapons recognition) was issued to the units by Director General Intelligence NDHQ, and was found useful for the lower ranks but insufficient for the officers and senior NCMs. Unfortunately, it was not received until after the preparation phase. Only limited country briefs were available. Much of the units' senior leadership time was taken up in staff work to garner or create training resources.

d. The units did not have all the equipment at their training base that they would be using in the UN base, which meant that some personnel operated some equipment for the first time after arriving in theatre (e.g., laser sights, global positioning system, CODAN radio, remotes for the NODLR).

e. The units were not authorized to communicate directly with the in-theatre units they were replacing; in fact, policy forbids this liaison. (The study team suspects that the units were forced to work around this policy and used *ad hoc* direct communication to gather valuable and necessary information.)

f. UN rules of engagement are of little use to the soldier in the form in which they are now produced, that is, highly politicized, and frequently involving lengthy decision-tree type rules. They are valuable to the commanding officer, officers commanding, and platoon commanders, but require each unit to produce its own "soldier" version. As stated in the RCD Post Operation Report: "Our soldiers are not lawyers; they need a simple document that leaves no doubt as to what force level they can use in the performance of their duties."

g. There was difficulty in allowing non-medical personnel to train in combat first aid, self-administered first aid, intravenous training, and trauma training (as opposed to St. John Ambulance type of first aid).

Both units were directed to devote considerable training time to meet collective, sub-unit live fire standards, even though one of the units is an infantry unit. One unit had great difficulty (the other had somewhat less difficulty) prioritizing the training requirements and balancing the mandated "risk reduction" training (individual battle task standards, collective training, mine awareness, etc.) with the non-traditional training (mission-specific details, stress management, negotiation, etc.). Emphasis was, of course, given to the former, leaving very little time for the latter. One unit spent 48 days training in the individual training phase and 15 days in the collective phase. Despite this, the units did not recommend extending the 90-day pre-deployment period (unless a change of role was required), fearing a "long list" mentality and "mission creep" (rather than trying to cover all aspects in a mediocre fashion, it would be better to cover the essentials well).

Both units found their confirmatory exercise very valuable. However, they expressed two concerns. First, during the confirmatory exercise, too much time was devoted to evaluating decisions made by senior leadership, at the expense of training junior leadership and soldiers. More time could be devoted to the latter if senior leadership was given a prior exercise (perhaps a command post exercise or a computer-assisted simulation) without the troops.

Second, because of the limited time available for the exercise, a false sense of pace of operations was imparted to the soldiers. In order to provide as many scenarios as possible to the largest group as possible, the exercise designers provide a number of "incidents." Frequently, trainees would be required to move on to a second incident before successfully completing the first one in a realistic time frame. For example, a negotiation situation, which, in reality, could take hours or days, might take minutes in the exercise. Training on the peacekeepers' need for patience tended to get lost in fast-moving exercises.

The units expected the "system" to produce more than it does. In general, there were few problems with the general military training other than an unclear ammunition allocation. However, the mission-specific training was, for the most part, *ad hoc* and impromptu, with little outside help being provided. Some of this latter training — e.g., negotiation, conflict resolution, knowledge of theatre — is critical to mission success. Post-operational reports, standing operating procedures, and contact with in-theatre units constituted the main source of information for the units. This works for rotational units, but obviously does not work for new missions: other procedures need to be developed.

The time period for pre-deployment overall training needs careful examination. Time required will vary from unit to unit depending on a number of circumstances, such as the mission and concept of operations, the start state of training, the desired end state of training, and whether a role change is involved (e.g., an armoured unit in an infantry role), and the risks. These factors, plus the overall time allocated, will dictate the balance to be struck between general-purpose combat training and mission-specific training. This examination must be carried out very carefully to avoid "tick-in-the-box" training and to allow adequate mission-specific training.

Canadian Forces Command and Staff College

The focus for this educational institution is war fighting at the strategic and operational level. It is a joint institution (army, air force, and navy), the student body being made up of senior majors and new lieutenant-colonels, including some foreign students. The main course is 45 weeks long. Some of the core curriculum, e.g., ethics, combat stress, and crisis management, is transferable to peacekeeping operations. Instruction ADM(PER) 7/89, dated May 30, 1989, constitutes the governing document for the current Command and Staff Course, and gives little specific guidance on peacekeeping training.

The college conducts about three lectures related to peacekeeping, one discussion period, and a one-and-a-half day exercise (FRIENDLY LANCE), which follows the progressive steps of peace restoration. Little consideration is given to peacekeeping training or to UN civilian police, and civilians in peacekeeping theatres (UN staff, non-governmental organizations, local civil society and government, etc.).

Primarily on its own initiative, the college is incorporating some peacekeeping training into its core programs, but the study team feels this could be increased, relative to the 45-week course, and better coordinated by the Officer Professional Development Council.

Canadian Land Force Command and Staff College

The training establishment focuses on war fighting at the tactical level, with some operational-level work. Two courses concern peacekeeping: the Canadian Land Force Staff Course (CLFSC, 20 weeks long) and the Commanding Officers Tactics Course (COTC, two weeks long).

The CLFSC student body is made up of senior captains, primarily army, with a few navy and air force students, including some foreign students. A large percentage of the student population has been on at least one UN tour.

Some core programs are transferable to peacekeeping (e.g., ethics, law of armed conflict). As part of its "Operations in a Unique Environment" series, the college conducts a two-day package on peacekeeping. This package includes:

a. A 100-plus page précis produced in May 1995, which covers such areas as the United Nations, Canadian policy and record, mounting a peacekeeping operation, humanitarian operations, training for peacekeeping operations, and the future of peacekeeping;
b. Guest lectures from former Force Commanders, NDHQ J3 staff and former UN battalion commanders (four periods);
c. A discussion period of four sessions, which includes student (UN veterans) presentations of various missions; and
d. A one-day exercise (OLIVE BRANCH), which includes the requirement to propose training guidelines and an outline training schedule.

The two-week Commanding Officers Tactics Course includes a one-day package covering areas such as UN organization and capabilities, DND/CF policies, preparation and training for peacekeeping operations, plus a panel discussion with a mission commander, a UNHCR representative, and commanding officers with UN experience.

The study team found the peacekeeping précis issued to the college students to be a good background document. It states:

Geopolitical briefings — Study of Mandate and Missions. Some aspect of these briefings may not be relevant to all personnel. Certainly officer observers should attend all of them. For peacekeeping forces the complete section may be relevant for selected officers and senior NCOs. Some study of the mandate/mission will be necessary for all ranks. The following should be considered:

a. Geopolitical briefings (geography, history, economy, political system, Defence Forces and Internal Security Forces, internal influences including religion, militias, revolutionary movements, external influences, and culture and customs.
b. The mandate and mission is central to all PKOs. As such it must be dealt with during pre-deployment training. How deeply one needs to go into mandate and mission study will be a selective decision by those responsible for training. In general, senior unit and staff officers should familiarize themselves thoroughly with the subject. At the other end of the scale, a simple explanation of the military mission, with discussion and a question-and-answer session should suffice.[36]

The college has a short, but concentrated, peacekeeping training package. Any expansion of this would have to be at the expense of some "war fighting" program. Without having conducted a curriculum review, the study team believes there is scope for increasing the time spent on this subject, if in no other way than by realigning some of the current core subjects towards peacekeeping scenarios, e.g., ethics, estimates of the situation, training development, etc.

Army Lessons Learned Centre and Peace Support Training Centre

The Army Lessons Learned Centre (ALLC) is being relocated and reorganized in Kingston, and has a small staff (four officers), whose role is the collection, analysis, and dissemination of lessons learned from Canadian and allied operational and training experiences. The centre will use electronic/all source data collection, primarily a preformatted post-operational report. It will also use temporary mission teams that will develop a collection plan (a series of detailed questions) for a deploying unit. The questions will form part of the unit's post-operational report. Issues will be analyzed to see if changes must be made to army doctrine, structure, equipment, or standing operating procedures. Dissemination of information will be by means of a text retrieval data bank and a quarterly lessons learned newsletter aimed at junior leaders.

A Peace Support Training Centre is also to be set up in Kingston. Initially slated to be staffed by September 1995 and to commence training by October 1995, those milestones have been delayed until 1996. Hopefully, the delay does not indicate residual reluctance to move on a concept that dates from at least as far back as the 1993 DCDS directive.[37]

Like the ALLC, the Peace Support Training Centre will have only a small staff, that is, one lieutenant-colonel, one major, two captains, and one sergeant, and will co-locate with the ALLC. The training centre's mission is to "provide training assistance in UN and peace support mission specific subjects for formed units and individual CF personnel posted to various missions."[38] It will "provide a nucleus of expertise within the CF responsible for the development of peace support techniques based on Lessons Learned, training methodology, training standards, and the provision of training and training support."[39]

Among other responsibilities, the Peace Support Training Centre will be tasked to "coordinate the development of peace support training and standards, standing operating procedures, training manuals and packages, *aide memoires* and bulletins for the CF."[40] It will also be tasked to coordinate out-service or civilian instruction for CF units conducting mission-specific training.

As a unit, the ALLC has huge potential. For the army, at least, it provides a focal point (a "corporate memory"), and a link between post-operational reports or after-action reports, with doctrine and training. The study team believes there are two potential problems, however. The ALLC appears to be under-resourced, with only four military personnel. The ALLC also appears to have multi-Command and national applicability and potential, which will be missed if it merely serves the army. This will be particularly true if the ALLC colocates and to a degree merges with the Peace Support Training Centre.

The Peace Support Training Centre has even more potential, as it is the mechanism for the "development of peace support techniques based on Lessons Learned," and the vehicle for transferring these skills through training to CF units and individuals deploying on peacekeeping. Its potential also seems threatened by being under-resourced, with only five staff.

Royal Military College

This educational institution trains about 25 per cent of new officers for the Canadian Forces. The bulk of the students' study time is spent on academic "civilian" subjects directed by the academic wing, with limited time allocated to the military wing, which directs military subjects *per se*. The college teaches some general military subjects, e.g., leadership theories, codes of conduct, military ethics, law of armed conflict, etc., that are transferable to peacekeeping operations. Aside from this officership training, no peacekeeping subjects are taught to undergraduates.

The college military wing is not mandated to provide peacekeeping training *per se*. Although this is probably appropriate, there is scope for the college to use peacekeeping scenarios in its core military training programs (e.g., military professionalism, military ethics, decision-making, cultural relativism).

Canadian Forces Recruit School

The school provides an eight-week course for new-entry recruits of all services. The subjects are very basic (drill, physical fitness, service rifle,

CF policy awareness, law of armed conflict, etc.). The school does not teach any subjects related to peacekeeping *per se*.

Scope does not appear to exist for including peacekeeping training in basic recruit training, other than an enhanced law of armed conflict portion. Such a portion could incorporate an "attitude" or "approach" to peacekeeping, as discussed in detail in Chapter 3.

The Royal Canadian Regiment Battle School

The school teaches various levels of courses, ranging from those for recruits arriving direct from the Canadian Forces Recruit School, to specialized courses on behalf of its area headquarters. No peacekeeping training is given to recruits. If so tasked by area headquarters, the school will conduct courses such as preparation for Operation MANDARIN,[41] which covers specific-to-theatre and general UN operations. The area headquarters mandated courses can easily be tailored to peacekeeping operations; many existing courses are currently being refocused in this direction.

There is some scope to use peacekeeping scenarios in recruit training at this more advanced level, particularly in such subjects as law of armed conflict, control of weapon fire, rules of engagement, field craft, etc.

Office of the Judge Advocate General

The Judge Advocate General (JAG) has a Directorate of Law/Training. This newly created directorate assumed the training function from another directorate, where operational priorities had not allowed for appropriate levels of effort in the training field. JAG is seen as having a role both in law of armed conflict[42] training and in related legal aspects of peacekeeping.

JAG provides a single annual course on the law of armed conflict, with an average of 40 participants, 30 of them Canadian military. It runs for one week and largely focuses on legal aspects, albeit using operations as context. JAG has also recently developed and provided a one-week course on operational law, which has a broader scope than the law of armed conflict course, and is primarily for lawyers.

JAG has a small library of related source materials, plus various posters and pamphlets on the law of armed conflict. These include a short publication designed for general readership, entitled *You and the Law of War*, and a lengthy draft publication for legal officers, senior commanders with their staffs, and other Canadian Forces members needing detailed background in applicable theory and policy, entitled *CF Law of Armed Conflict Manual.* Unfortunately, the manual is out of date in certain sections (it was drafted in 1986), and the chapter on UN peacekeeping is extremely short and general.

While JAG should be the responsibility centre for training on the law of armed conflict, it has not been given that mandate, or the necessary resources (discussed further in Chapter 3). While over 43 CF training courses at least mention the law of armed conflict, JAG has little idea as to the content and methodology of that training. Beyond the annual course offered, JAG's involvement in such training is *ad hoc*. It appears that frequently the law of armed conflict is taught without any contact with JAG, or, at best, with irregular contact. For example, in the Senior Leaders Course in Borden, if a JAG staff member is unavailable, then a member of directing staff is substituted to deliver the module. In light of the complicated nature of the law of armed conflict, this solution appears to have a number of flaws.

More time must be devoted to teaching the law of armed conflict, and better use should be made of time currently allocated to the teaching methodology applied to the subject. Such specialized training requires specialized trainers, who understand both the legal and operational facets of the law of armed conflict, and know how best to transfer this knowledge to particular audiences. JAG expressed concern about the apparent pattern of cursory and standard package delivery by non subject matter experts. Assuming one trains operators as subject matter experts, this does not conflict with a concern regularly repeated by those interviewed at the operational level, who stressed that training in the law of armed conflict should be delivered in large part by operators. They also stressed that operational application of the law of armed conflict should be paramount.

Study Team General Observations

The Department of National Defence and the Canadian Forces do not have a coherent, integrated, organic peacekeeping training program. Preparation training for peacekeeping is not coordinated by any single military, civilian, or academic agency. As such, peacekeeping training, as opposed to general-purpose combat training, is only marginally fitted into the normal Canadian Forces training matrix.

Peacekeeping training within the Canadian Forces is largely concentrated in the 90-day pre-deployment training period of a unit warned for a

UN mission. Otherwise, there is little training time devoted by units to specific peacekeeping training, e.g., in their annual training cycles.

With regard to individual training, little, if any, specific peacekeeping training is being conducted at any of the basic training institutions. The only peacekeeping training the study team found in the advanced individual training category was some limited training of officers at the Canadian Forces Command and Staff College, Canadian Forces Land Command and Staff College and professional development levels. In Land Force refresher training there are no direct peacekeeping individual battle tasks standards. With regard to collective training, almost no peacekeeping training exists in the generic annual training cycles of the units.

In summary, *there is no direct peacekeeping training at the basic level, very little training at the individual level, and almost none in the generic annual training cycles of the units. It is concentrated in the 90-day pre-deployment training period of a unit warned for a UN mission.*

It is the study team's view that this situation is no longer appropriate for the new era and the new peacekeeping paradigm referred to by the Senate Standing Committee. Increasingly, the majority in the Department of National Defence and the Canadian Forces are coming to the same conclusion. The recommendations offered in Chapter 3 are designed to provide suggestions on how non-traditional training for peacekeeping could be put into effect.

CHAPTER THREE

Recommendations on Training Requirements

INTRODUCTION

The study team was very conscious of the fact that peacekeeping training, indeed any Canadian Forces training, must be viewed in context, giving due consideration to all the roles and tasks of the Canadian Forces. This fact is perhaps best summed up in the 1994 Defence White Paper statements:

The Government has concluded that the maintenance of multi-purpose, combat capable forces is in the national interest.[1]

Canada needs armed forces that are able to operate with the modern forces maintained by our Allies and like-minded nations against a capable opponent — that is, able to fight alongside the best, against the best.[2]

This study is based upon the belief that a good peacekeeper is also a good combat trained soldier, and that general combat readiness provides critical skills for Canadian peacekeepers.[3] That belief fits well with the study team's conclusions that UN peacekeeping in the 1990s requires additional non-traditional military skills, as well as a reorientation of some general combat readiness attitudes in order to be more effective in peacekeeping operations. Such additional training and reorientation when involved in peacekeeping operations should leave the traditional military or warrior skills intact.[4]

A joint study by the Department of National Defence and the Department of Foreign Affairs has recently confirmed that "traditional military training should be modified to include the unique tasks of peace missions because armed forces are now called upon to perform a wide variety of tasks to help resolve conflicts, as well as to be prepared for combat."[5]

RECOMMENDATION 1. *It is recommended that the Canadian Forces overall training philosophy be amended so that general-purpose combat training, while remaining the foundation of training policy, is supplemented by additional non-traditional military training geared specifically for UN peacekeeping operations.*

Those interviewed frequently commented on the tension between general-purpose combat training and training for peacekeeping. One example was that of coming upon a roadblock. A war fighting situation and general-purpose combat training would invariably call for a much different response than would normally be the case in peacekeeping, e.g., possibly getting down from one's vehicle and having a slivovitz with the defenders as a preamble to negotiating passage. Another example often cited concerned weapons training. General-purpose combat training would train soldiers to aim at the centre of the visible mass for a kill or a first-round hit. For peacekeeping, in keeping with the normal use of minimal force, weapons use would range across a broader range of weapons response, including the use of warning shots.

The question is not whether peacekeeping should be taught, but how it should be layered over general-purpose combat training, so that core traditional military skills are not lost or made too difficult to regenerate when required.

The challenge ahead for the Canadian Forces is how to meet increasing requirements with a decreasing military budget and fewer personnel. Recent budget cuts have already necessitated a reduction in training. Future demands on the system will be many, and prioritization will be difficult. It is unlikely that there will be additional resources for a large-scale peacekeeping training scheme, and cost and time objections to any additional training for the Canadian Forces are valid concerns.

Nevertheless, the Canadian Forces have continuously demonstrated that their greatest resource — servicewomen and servicemen — are able to use great initiative and adapt to new conditions. A priority-setting exercise that includes interactive and iterative dialogue between the political level and the senior military staff must be put in place in order to enunciate a clear overall training policy. If the demand for peacekeeping operations continues at its current level, constituting the principal operational activity of the Canadian Forces in general and the army in particular, then there will be a pressing need for increased peacekeeping-specific training at all stages of unit and individual development.

47 Recommendations on Training Requirements

The following *Recommendations on Training Requirements* elaborate upon this requirement in three substantive sections:

What Skills recommends the type of peacekeeping skills, and thus training, required for the new breed of Canadian military peacekeepers.

Training: For Whom, When, Where provides recommendations on where such training could be given.

Other Sources of Training Practice and Guidance recommends some alternative sources of peacekeeping training methodology, material, and course content.

WHAT SKILLS

Skills for UN Peacekeeping

It may be useful here to repeat some comments from Chapter 1 that discuss the content of peacekeeping training, i.e., non-traditional military training. The study team's point of departure was to look at the lengthy list of skills required to meet possible operational demands of a UN peacekeeping operation. While not completely unique to peacekeeping operations, these skills are at least substantively different from the skills and attitudes usually taught to Canadian military. They include:

Strategic/Political:

1. Operational limits imposed by UN mandates;
2. Operational uncertainties resulting from imprecise UN mandates and rules of engagement;
3. Working under UN field operational control while remaining under the command of Canadian authorities;

Operational:

4. Different, and at times confusing or minimal, UN standing operating procedures;
5. Working with and alongside non-NATO military contingents and CIVPOL;[6]

6. Working alongside large non-UN international agencies, e.g., NGOs, ICRC;[7]
7. Working alongside major UN agencies, e.g., UNHCR, WFP, UNDP;[8]
8. Working alongside other substantive UN civilian components such as a human rights division or a legal division;

Theatre Environment:

9. Having to deal directly with foreign populations and authorities, particularly at the tactical or community level;
10. Having to deal with armed parties to a conflict as non-enemy albeit at times with a very real possibility that they could become the enemy for reasons of self-protection or UN Charter Chapter VII art. 42 use of force;

Specialized Training:

11. Conflict mediation and resolution;
12. Dealing with issues of human rights violations, e.g., monitoring human rights violations;
13. Dealing with issues of humanitarian assistance;
14. Playing a role in post-conflict rehabilitation;
15. Critical incident stress management; and
16. Operational knowledge of the law of armed conflict.

For the purpose of this study, non-traditional military training refers to the training of skills to deal with the peacekeeping operational criteria mentioned above. In other words, it is training for other than general combat roles.

While a detailed analysis of non-traditional military training requirements is well beyond the scope of this study, the UN peacekeeping training needs are discussed below in more detail. The four themes that cover those needs — *Strategic/Political, Operational, Theatre Environment,* and *Specialized Training* — are far from watertight, but such a division provides some indication of the pattern of peacekeeping training required, and may help in making decisions as to who within the CF should be receiving such training.

49 Recommendations on Training Requirements

Strategic

The first group of skills (#1, 2, and 3), labelled *strategic/political,* consists of a number of political and strategic issues such as mandate, command and control, concept of operations, and rules of engagement.

In the beginning, when a UN mission is being contemplated, there is a very complex decision-making matrix of the UN Security Council, the UN Secretary-General, the UN Secretariat, and the political/military structure of troop contributing nations. While Canada is trying to influence the United Nations section of this decision-making loop,[9] there is definite scope to examine how Canada functions internally at the strategic level. A look at the make-up and responsibilities of such bodies and organizations as the Cabinet, the Departments of Foreign Affairs and National Defence, National Defence Headquarters, and Canada's Ambassador to the UN, reveals a potential for overlap and omission in the manner in which Canada agrees to or sets these strategic decisions.

There are two obvious ways of minimizing the potential for such eventualities. The first is to organize the political/military interface more effectively so that it is a collaborative process, but recommendations on this are clearly beyond this report's mandate. The second, which is germane to this study, is to ensure that sufficient military and civilian personnel are trained at this strategic level to fill the key positions in DFAIT, DND, and UN Headquarters.

RECOMMENDATION 2. *It is recommended that military and civilian personnel selected for positions involving peacekeeping operations receive training (at the strategic level) on subjects such as UN decision-making, mandate formulation and interpretation, UN and national command and control mechanisms, and rules of engagement formulation and interpretation.*

Another issue that officers and senior NCMs need to be aware of is the different concept of command that exists in a UN operation. Troop contributing nations tend to have varying degrees of trust in UN operational and tactical command, and thus there is a fair amount of "shadow" control from national capitals. In addition, and depending on the force commander and their staff, command tends to be looser than would be found in a Canadian or NATO context. Thus commanders at various levels tend to have greater operational discretion, albeit within the mission mandate and rules of engagement.

RECOMMENDATION 3. *It is recommended that doctrine be developed on the concept of "unity of effort" in UN operations, e.g., operating within the normally loose and poorly defined UN chains of command which frequently involve civilian organizations, and that this doctrine be practised by the Canadian Forces during some of their collective training exercises.*

One key aspect of all Chapter VI peacekeeping, and even most Chapter VII peacekeeping, is that of dealing with armed parties to the conflict as nonenemy. There is a need for Canadian military to have more extensive training in flexible response to conflict situations, to reflect both UN operational and mandate requirements. Included in this is the need to be aware of techniques of both graduated escalation to peace enforcement, and de-escalation to steady state peacekeeping and eventual withdrawal. This is discussed further in the section on the relevance of Canadian police training to peacekeeping training.

Operational

The second group of skills (4, 5, 6, 7, and 8), labelled *operational,* deals with the many operating differences that would be faced by Canadian military in peacekeeping as opposed to war fighting in the NORAD or NATO context. This includes the often very different UN standing operating procedures, administrative, and logistics mechanisms. Increased training or preparation in this skills group will increase the ability of Canadian units and individuals to operate within UN field operations.

RECOMMENDATION 4. *It is recommended that Canadian military receive training on the unique character of UN operations in such areas as its standing operating procedures, administration, logistics, and terminology.*

UN operations also involve a wide variety of military and civilian partners. There are now over 80 countries providing military peacekeepers. Even more varied and complicated is the variety of civilian field partners, both UN and non-UN. This ranges from UN political staff and CIVPOL, to a wide variety of organizations or UN field mission components in areas such as humanitarian aid, developmental assistance, human rights, and media. UN military invariably have a preponderance of resources and personnel, and are more geographically dispersed yet with a unified command and communications structure. For this reason, they often end up playing an informal coordinating role, as witnessed by the increasingly common creation, initiated by the military, of civilian-military operations centres (CMOCs) or similar mechanisms for civilian-military coordination and inter-agency operations.

In addition, Canada is often asked to provide individuals for UN field headquarters, or specialized units such as communications or logistics. These individuals or units either command or coordinate action by numerous national military contingents, and are frequently involved in civil-military coordination. Thus, their awareness of the cultural and functional variations among national contingents and other peacekeeping partners is of key importance in achieving unity of effort in the mission.

RECOMMENDATION 5. *It is recommended that Canadian military receive training on dealing with other military and civilian field partners, so as to increase Canadian ability to play a role in enhancing unity of effort by all civilian-military components of a UN field operation.*

Theatre Environment

The third group of peacekeeping skills (9 and 10), labelled *theatre environment,* deals with the rather unique relationship of UN peacekeepers to the local population and the parties to the conflict. This is not only a different skill set from general-purpose combat training, but includes a dramatically different attitude or approach. This need for a different operations attitude is much harder to address successfully in training, for in many ways it cannot merely be an overlay of the traditional military approach to combat.

Most aspects of training in aid to the civil power, as presently received by Canadian military, are relevant to peacekeeping, e.g., crowd control. This type of training should be examined as to the need to tailor or expand it for peacekeeping needs.

Communication with the local population and parties to the conflict is absolutely essential for the success of a UN peacekeeping mission. Canadians who have not worked abroad often assume that regardless of the spoken language, people will communicate their intentions and reactions in the same way that Canadians do. Experience has shown that this is often not the case, and ignoring such differences in cultural and behavioural means of communication can be dangerous.

A lack of knowledge of the cultural context of the mission theatre can and has led to a misreading of individuals' and groups' intent, and a failure to recognize opportunities for concessions, agreements, cease-fires, or other steps forward in peacekeeping objectives. For peacekeepers, a lack of understanding of the cultural environment within which they are both working and living will also increase feelings of isolation and stress, and contribute to the perception of surrounding populations as "them," and thus probable antagonists. This in turn can lead to inappropriate behaviour by peacekeepers and decrease peacekeeping effectiveness.

Mission-specific training must alert Canadian peacekeepers to the issue of cultural context, and to the unique dynamics of communications in the mission assignment region. Three types of preparation can contribute to this training.

First, an overview of cultural behaviour patterns should be centrally researched and prepared, and proactively distributed to all members of an assigned unit. Second, a briefing which utilizes cultural training specialists and prepared representatives of the populations in question should take place in the 90-day pre-deployment period. Such briefings have taken place with good results recently for peacekeepers going to Rwanda and Haiti, but were initiated by the assigned units, apparently without central guidance or sourcing support.

Third, a structured capacity should be created for collecting lessons learned about dealing with local populations in the field, and for passing these on as an integral part of the training and preparation for units deploying. All of the above preparation should be varied from mission to mission — a complexity that calls for a central armed forces responsibility centre.

RECOMMENDATION 6. *It is recommended that a guide to the mission's cultural behaviour context, including factors such as religion where significant, be prepared centrally and distributed to all individuals or unit members deploying on mission. This should be carried out by a central responsibility centre which is also tasked with collecting and articulating lessons learned for subsequent guides and troop rotations.*

RECOMMENDATION 7. *It is recommended that a training session on dealing with the local population, involving nationals from the mission area or subject matter experts, be an element of each unit's pre-deployment preparation, and that guidance and sourcing support for such training be provided by a central responsibility centre.*

RECOMMENDATION 8. *It is recommended that, as much as possible, subjects such as country briefs, population details, ethnic characteristics, culture, etc., be largely taught by experts or unit officers rather than by intelligence cells.*

Frequently, the language spoken by most of a local population will not be English or French. Often forming key command components of a UN peacekeeping mission, and equally often operating away from other UN resources, Canadian peacekeepers must view language/communication capacity as a core skill. The effective use of translators through an understanding of their capacities and limitations is a unique skill that can and should be taught to those using them. In addition, foreign-language training should be considered for a small number of Canadian peacekeepers,[10] to provide that communication capacity for critical situations. They would have the supplementary role of monitoring the use of translators.

Although it is unrealistic to expect that a great number of front-line personnel will pick up much language fluency, the fact is that those in most direct daily contact are unlikely to have the support of translators on an ongoing basis. Any amount of progress in language skill will increase their comfort and effectiveness: whatever support personnel can use in learning the language should be made available as requested. Self-taught language guides and language classes for voluntary attendance should be made available, as they are occasionally now, for in-theatre training.

RECOMMENDATION 9. *It is recommended that at least one individual per battalion-size unit deployed on peacekeeping be sufficiently trained to speak the predominant local language(s), and that other peacekeepers using translators be trained on their capacities and limitations. Support for further self-directed language learning should also be provided in the field.*

Specialized Training

Finally, the fourth group of skills (11, 12, 13, 14, 15, and 16), labelled as *specialized training,* deals with the need for the military to at least facilitate and at times be substantively involved in issues of conflict resolution, human rights, humanitarian assistance, post-conflict rehabilitation, stress management, and the law of armed conflict.

These tasks sometimes result from the imperatives of the situation at hand, especially where military peacekeepers may be the only, or the most predominant, UN presence. Conflict negotiation and management is perhaps the best example of this, as is dealing with particularly egregious violations of fundamental human rights. At other times, various roles will be ancillary mandates for UN military to assist other UN and non-UN field partners that have the primary responsibility for human rights, refugees, humanitarian assistance, development assistance, etc.

These special skills are, by their nature, relatively complicated and require more intensive training. The confusion and multiplicity of tasks facing those preparing to deploy militates against sufficient time or ability to concentrate on such training.[11] These special skills should be predominantly taught as part of regular training, in both CF training establishments and in regular unit training.

Most Canadian peacekeepers interviewed referred to low-level conflict resolution. Without exception, those mentioning it felt that they had received insufficient or no training in mediation and conflict resolution. Several referred to other national peacekeepers, such as New Zealanders, who appeared to have a more trained and professional approach to field negotiations.[12] In light of the nature of peacekeeping, where small patrols can suddenly be faced with an incipient crisis, there is a need for almost all ranks to have some capacity in this regard. Even lower ranks should have some techniques for calming and stabilizing a situation until more highly trained mediators and negotiators (senior NCMs and officers) arrive.

RECOMMENDATION 10. *It is recommended that low-level conflict mediation be taught to all junior NCMs, and that more refined mediation and conflict resolution skills be taught to senior NCMs and officers. This training should largely occur as part of regular professional and unit training, but should be customized during pre-deployment refresher training to address the particular cultural/political environment of the theatre of operations.*

Increasingly, it is felt that UN peacekeepers have a human rights role. The overwhelming mission-wide presence of military, relative to other UN personnel, is the first argument in favour of this role. This has been borne out in a number of different situations where UN military themselves have initiated action, as they were not prepared to stand by and watch fundamental human rights being violated. This predilection is particularly true for contingents from countries with strong internal human rights protections and traditions.

Aside from observing and reporting, one of the military's advantages in assessing human rights violations is their greater awareness of command responsibility within the hierarchy of the parties to the conflict. This facilitates assigning accountability for violations carried out by combatants. The military also have quite specific skills, such as crater analysis and crack-thump training, which allows them to conclude with greater certainty issues such as what is being fired, from where, by whom, and with what intent. Monitoring and assessing human rights violations is also becoming recognized as an important intelligence-gathering mechanism to inform tactical peacekeeping decisions of both military peacekeepers as well as the larger UN mission.

Five of the most recent UN peacekeeping missions have had a distinct human rights division, with which other UN mission components, such as the military, were expected to cooperate. This recent and dramatic growth in human rights tasks in UN operations has had operational successes and will only increase. The resulting human rights role for military peacekeepers *per se*, requires specialized training for Canadian military. Basic human rights training, on what violations to look for or recognize, and how to report them, is required down to the lowest operational unit size. At times this could be a section commanded by a senior corporal or sergeant, or a single military observer. More specialized training is needed for senior NCMs, lieutenants, captains and majors, in order to be able to assess human rights intelligence and make informed decisions on appropriate action aside from mere observation and reporting.

RECOMMENDATION 11. *It is recommended that the Canadian Forces train various military as specialists in human rights monitoring and reporting, both to work with those UN field staff coordinating human rights promotion and protection, and also to interpret human rights intelligence to guide Canadian peacekeeping tactical decisions.*

Humanitarian assistance is very much the responsibility of both UN agencies (e.g., UNHCR, WFP, UNDP, UNICEF, DHA), bilateral government agencies (e.g., CIDA, USAID), and international NGOs (Oxfam, CARE, MSF). However, military peacekeepers invariably play an important supportive role, ranging from providing security for those responsible for providing assistance, to actually delivering assistance in unique emergency situations where the usual agencies are not present. Although this ties in with the operational skills mentioned above, on how to work alongside a broad variety of non-military partners, the potential involvement of Canadian military in the humanitarian assistance field is sufficiently large as to merit the training of CF military specialists in humanitarian assistance.

An example of this increased military involvement in humanitarian assistance is the evolution of UNHCR's concept of humanitarian services packages to enhance humanitarian rapid reaction. Broadly speaking, UNHCR puts out requests for various services, as it did for the provision of water at the Goma camp for Rwandan refugees, or running the Kigali airport, including air-traffic control, security, and maintenance. Countries or organizations then select which services package they wish to undertake. Invariably, a limited number of militaries, including Canada's, have the capacity to meet such rapid reaction goals. A reflection of such an evolution is the seconding, by various militaries, of officers to UNHCR headquarters in Geneva.

RECOMMENDATION 12. *It is recommended that DND train various military officers as specialists in humanitarian assistance, both to facilitate military field support for the traditional agencies providing such assistance, and also to advise any Canadian peacekeeping units that might be specifically tasked to provide humanitarian assistance.*

Post-conflict rehabilitation, like humanitarian assistance, is invariably the purview of specialized agencies such as UNDP. However, this rather broad category of activity was referred to in the 1994 Defence White Paper: "The rehabilitation of areas that have been the scene of armed conflict represents an important contribution that the training, skills, and equipment of our armed forces can make to security abroad. Past instances of such contributions include the provision of humanitarian relief supplies and the use of engineers to rebuild infrastructure and remove land mines. ...training refugees to recognize and disarm land mines. These activities can make an invaluable contribution in building more durable peace, and the Government will explore ways in which the Canadian Forces can contribute further."[13] Often such activities stem from the field presence of Canadian military capacity for such rehabilitation. When there is steady state peacekeeping and no operational need for this capacity, it can usefully be turned to rehabilitation activities.

RECOMMENDATION 13. *It is recommended that the Canadian Forces train various military officers, particularly those with engineer and support roles, as specialists in post-conflict rehabilitation so as to maximize the contribution of the training, skills and equipment of certain peacekeeping units or sub-units such as field engineers that might be present yet relatively underutilized during various stages of a peacekeeping mission.*

Similarly, such specialists could advise units that are specifically deployed to effect post-conflict rehabilitation.

Critical incident stress management is already a part of general combat readiness, and as such is a valuable preparation for peacekeeping. However, there are some unique situations faced by peacekeepers that merit being added to existing stress management training. The possibility of having to deal with large refugee flows, human rights violations such as ethnic cleansing, and humanitarian catastrophes up to and including starvation, can create added pressures on peacekeepers who are expected to involve themselves in addressing such situations. Even more stressful are the situations when the UN and its peacekeepers are constrained by political or resource limitations, and can do little or nothing to alleviate suffering. Critical incident stress management for peacekeepers must be tailored to deal with these and other rather unique peacekeeping stresses.

RECOMMENDATION 14. *It is recommended that critical incident stress management be emphasized as a key component of general combat readiness, and that training to manage critical incident stress be augmented to deal with incidents relatively unique to peacekeeping operations.*

Training in the operational knowledge of the law of armed conflict[14] has been categorized as specialist training, but this should not imply that it be restricted to a small group of specialists. In fact, this specialized knowledge must be broadly taught throughout the Canadian Forces. "[The] success or failure of peacekeeping missions rests to a great degree on the local population's perceptions of the peacekeepers, so the tactical and strategic consequences of violating the laws of war in peacekeeping missions could be greater than during combat."[15]

Since Korea, Canadian troops have had infrequent involvement in combat situations where Canadian troops would have truly faced the ethical and legal challenges posed by the law of armed conflict. Cyprus-type UN operations largely failed to pose such complicated issues, and understandably this issue was not seen as a danger for Canadian troops in classic peacekeeping operations. Somalia-type UN operations have changed this perception, but as Chapter 2 of the study discusses, widespread CF training to date in the law of armed conflict has lagged behind.[16] Substantive training has been largely restricted to military lawyers, primarily short legal lectures for officers, and minimal operational training to rank and file. In addition, many of the operational-level personnel interviewed remarked that dry legal lectures by military lawyers were not particularly helpful. They felt that, particularly for NCMs, training by their own officers and warrant officers would be much more effective.

The International Committee of the Red Cross has recently developed new training modules and materials, in a renewed attempt to set out the training in operational terms, and make more relevant the legalistic format of the laws of armed conflict. Although the Office of the Judge Advocate General is aware of, and has most of these materials, it has lacked both the resources and the clear mandate to expand, update, and oversee training now provided to most officers and to all other ranks. JAG's role in this is discussed at length in the next section.

Quite aside from specialist training in human rights, as mentioned above, there is a need to address human rights operational standards for peacekeepers,[17] which are increasingly taught to UN CIVPOL and logically should be taught to UN military as they undertake what are often low-level conflict functions.

RECOMMENDATION 15. *It is recommended that teaching the law of armed conflict become much more prevalent and extensive, and that it be taught as an operations subject with clear field applicability as opposed to a legal skill. It is important that international human rights law and standards, particularly as refined by the UN for low-level conflict CIVPOL functions, be incorporated into such training.*

What Other Peacekeeping Skills

As a qualification set out at the beginning of this section, it was recognized that a detailed elaboration of non-traditional military training is well beyond the scope of this study. It is incumbent upon DND to carry out a thorough analysis of the military's evolving tasks in peacekeeping, and to identify in depth and in detail just which skills can and should be imparted to some or all members of the CF and DND at large.

RECOMMENDATION 16. *It is recommended that J3 Peacekeeping, as the office of primary interest, create and chair a DND-wide working group to undertake the identification in depth and in detail of non-traditional military skills needed for peacekeeping.*

TRAINING: FOR WHOM, WHEN, WHERE

General

The first question in training is who should receive it: Officers? Senior NCMs? All NCMs? Civilians? Traditionally, military training has tended to reserve tactical "thinking" skills for officers, "trade" skills for NCMs, and has given little thought to training civilians. The Canadian Forces have modified this categorization in recent decades, but training still reflects the reality of relatively large and cohesive military units and sub-units that traditionally need very few people at the top taking the "larger" decisions.

Peacekeeping, however, results in smaller and smaller sub-units being deployed, often in relative isolation. Frequently, military peacekeeping involves individual or section-level action where junior officers and junior NCMs are faced with decision-making situations which they need to resolve in whole or in part by themselves. In addition, various civilian DND staff play critical roles in supporting Canadian peacekeepers. Their support functions will be far more effective if they are more cognizant of the issues and difficulties faced by Canadian peacekeepers.

As made clear in the introduction to Chapter 3, the Canadian Forces needs to develop a training philosophy and policy that embraces both general-purpose combat capability and nontraditional training. One of the most difficult decisions to be made is how to achieve that balance. However, it is inappropriate to direct all peacekeeping training only to those warned for a peacekeeping mission, and immediately prior to their deployment. There is a core of subjects (some already discussed and others discussed below), which can be developed for introduction throughout universal personal development.

RECOMMENDATION 17. *It is recommended that once DND has identified in greater detail the content of non-traditional military training for peacekeeping, that J3 Peacekeeping, as the office of primary interest, create and chair a working group which would include the Director of Military Training and Education (Directorate of Military Personnel), as a key office of collateral interest, to undertake the identification of which components of DND, officers, senior NCMs, all NCMs, and civilians need to receive non-traditional military training for peacekeeping.*

RECOMMENDATION 18. *It is recommended that the Canadian Forces develop a core program of non-traditional training that will be received by all components of the Forces, and those civilians of DND who are involved in these operations.*

The second question in training is deciding when it should take place. Some have suggested that mission-specific training just prior to deployment is largely sufficient. However, the number of potential peacekeeping training demands set out above do not fit easily within pre-deployment limits of time and space. For example, the length of a pre-deployment training period seems to be a variable, depending on a number of factors, such as type of mission, urgency to deploy, etc. It was clear to the study team that a 90-day mission-specific training period (stated by the CF as the desirable period) would not usually be extended because of the other taskings of the units. It was also clear that the requisite general-purpose combat training, departure administration, leave, etc., left little time for general peacekeeping training, let alone mission-specific peacekeeping training. Thus within the 90-day period, a delicate balance must be achieved between combat capability and mission-specific training.

Training standards should be developed which prioritize training into "must know," "should know," and "could know" subjects. Mission-specific training needs to be balanced between "risk" (e.g., mine awareness, weapon recognition) and other subjects such as theatre awareness, law of armed conflict, stress management, low-level conflict resolution, negotiation procedures, culture awareness, non-governmental organizations, media relations, etc. "Training packages" need to be prepared for these "non-traditional" subjects.

There is some room for reordering pre-deployment training priorities, and dramatically streamlining and making more efficient much of the training that does take place. This latter point is elaborated upon below in sections dealing with unit training and the proposed Peace Support Training Centre. It should also be noted here that most individuals deploying as individuals or small teams, and some units, are not provided with a 90-day mission-specific training period. This is particularly true for composite units or those units composed in part by augmentees, who often arrive at different points along the unit's preparation time line, thus hampering the unit training schedule and leaving gaps in the individuals' training.

RECOMMENDATION 19. *It is recommended that the pre-deployment training period should be at least 90 days. This may be reduced if the unit was on UN standby and may need to be increased if the unit is composite or has a lot of augmentees. Training for individuals is more a variable depending on the mission, but needs to be extended beyond the few days now spent on this training to a period of about 14 to 21 days (more for observers, less for staff officers).*

The relatively short time available for pre-deployment training, and all other predeployment activities, places serious limits on the amount of peacekeeping training that can be carried out during this lead time. The obvious solution is that pre-deployment peacekeeping training consist largely of refresher training, where previous peacekeeping training is reviewed and given a mission-specific "customization." This is very much akin to pre-deployment general-purpose combat training, which serves to review past training and emphasizes what is needed for the military task at hand. Therefore, there is a need for substantial periodic peacekeeping training both by units and by individuals as they move along their career paths.

RECOMMENDATION 20. *It is recommended that in light of the finite scope for the pre-deployment training period and the limits that imposes on nontraditional and mission-specific training, a core of peacekeeping subjects be taught in advance at regular stages in unit and individual training. These peacekeeping skills, as with general combat readiness, will be merely refreshed and refined during the annual and predeployment periods.*

The third question on training is where it should take place. There is the obvious attraction of having peacekeeping training as an integral part of existing military training mechanisms. This can result in more comprehensive training throughout the armed forces, and at all stages of an individual's military career. Arguably, this approach would also be more cost-effective than creating new training establishments. This does not contradict the need for a new facility to help units undertake mission-specific peacekeeping training, and the study team understands this to be the rationale behind the proposed Peace Support Training Centre (discussed below).

RECOMMENDATION 21. *It is recommended that non-traditional military peacekeeping training be an integral part of most existing military training mechanisms and establishments.*

Assuming recognition of the need for non-traditional military peacekeeping training as an integral part of most existing military training mechanisms and establishments, and that those establishments respond, a series of recommendations for training in a variety of specific locations is set out below.

National Defence Headquarters

National Defence Headquarters should be the focal point for peacekeeping policy and for setting joint peacekeeping training priorities. It appeared to the study team that no coherent flow of policy, doctrine, standards, priorities, resources and tasking exists. A clear policy document that enunciates the requirement for peacekeeping training and assigns responsibilities is needed. A lukewarm message as to the importance of non-traditional military training for peacekeeping will be reflected in the implementation of such training.

RECOMMENDATION 22. *It is recommended that NDHQ make it clear in both the wording and the spirit of training policy that Canadian peacekeepers in the 1990s require enhanced non-traditional military training for peacekeeping.*

NDHQ has recently authorized that a peacekeeping training staff be set up under Land Force Command. The study team notes that LFC is by far the major provider of troops for peacekeeping and supports this decision. However, the implementation leaves much to be desired. Three of the five positions necessary to create the LFC peacekeeping cell would come from the NDHQ J3 Branch, leaving only one officer in J3 to provide overall armed forces peacekeeping specialist advice, and training policy and doctrine oversight. *De facto*, this dissipates the central armed forces operational focus, and leaves peacekeeping training to other parochial staff branches of NDHQ.

RECOMMENDATION 23. *It is recommended that NDHQ create a single, central and joint peacekeeping training section within its organization with primacy amongst the staff matrix. This section would work closely with the soon-to-be-created Peace Support Training Centre to be set up under Land Force Command, but would not be replaced by it.*

The NDHQ peacekeeping training section would be responsible, among other tasks, for formulating training policy, assigning joint doctrine production responsibilities and obtaining national-level approval of that doctrine, approving peacekeeping standards recommended by subordinate commands, and assigning responsibilities for the production of training packages, standing operating procedures, *aide memoires*, etc. Early candidates for this latter requirement are packages on stress management, low-level conflict resolution, negotiation procedures, law of armed conflict, cultural awareness, and media relations.

As an adjunct to the aim of assisting units preparing for peacekeeping, NDHQ needs to look at its procedures for concurrent activity in relation to the formulation and dissemination of the mission, the concept of operations, the command and control arrangements, and the rules of engagement for new missions. It is recognized that the approval of these components may be a highly political and therefore a lengthy procedure, but it is also recognized that these are critical pieces of information needed by a unit training to deploy on a new mission.

RECOMMENDATION 24. *It is recommended that NDHQ examine methods of quickly disseminating operational information needed by units about to deploy, so as to allow them to effectively design and deliver their unit training.*

Perhaps as a result of a lack of a firm statement of requirement for peacekeeping training in the CF training philosophy, little follow-on institutionalized professional development training is being conducted.

RECOMMENDATION 25. *It is recommended that the Officer Professional Development Council examine the mandates given to CF staff colleges, military colleges and personnel sections with a view to formalizing peacekeeping training objectives for the various levels of an officer's professional development.*

Units that are training to deploy on a UN mission need training assistance from multiple sources. For units that are rotating into a mission, one of the obvious and principal sources of information would be the in-place unit.

RECOMMENDATION 26. *It is recommended that the policy of not having direct contact with in-place units be examined with a view to allowing replacement units to have contact with in-place units for training matters.*

Command Headquarters

Command headquarters, particularly Land Force Command, have major responsibilities in the formulation of peacekeeping doctrine and training standards, and in the setting of priorities and allocation of resources for training.

RECOMMENDATION 27. *It is recommended that the Commands institutionalize a flowing and coherent system of analysis of peacekeeping policy, the originating of peacekeeping doctrine (initially as a single service but inputting into joint, tri-service doctrine), and the creation of peacekeeping training standards.*

RECOMMENDATION 28. *It is recommended that the evolution of the Army Lessons Learned Centre and the creation of a Peace Support Training Centre at Land Force Command be pursued with vigour and that these centres be tied into the above system to provide both a corporate memory based on past experiences, and an input into future doctrine production. It is emphasized that the resourcing of these centres should not be at the expense of a national tri-service focal point.*

Like NDHQ, the Commands need to give clear guidance on the balance of training time between general-purpose combat training (e.g., individual battle task standards), generic peacekeeping training, and mission-specific peacekeeping training, and whether that training will be given in annual or refresher training, or pre-deployment training. It is the study team's view that pre-deployment training should concentrate less on battle tasks and more on peacekeeping. This assumes an otherwise satisfactory annual training program.

A possible approach to the pre-deployment training period might be to notionally split it into three time blocks. The first type of time block would be general-purpose combat training and would be a variable period depending on the state of training of the unit (e.g., has the unit completed annual refresher training or not, collective training or not, does it include augmentees, etc.), and its future mission (e.g., has there been an equipment or role change).

The second type of time block would be devoted to "peacekeeping training," both in the generic and in the mission-specific sense, and would be a more or less constant time period. The third type of time block would be for departure administration and leave, and it too would be a constant period. Thus, in all circumstances, short of no-notice situations, the deploying unit would be "guaranteed" mission-specific training time, regardless of its general combat readiness, which would be catered for separately. The time periods would not be separate and would overlap.

RECOMMENDATION 29. *It is recommended that, to consciously protect the time allocated to peacekeeping training, the Commands examine notionally splitting pre-deployment training into three overlapping blocks: general-purpose combat training; peacekeeping training; and departure administration.*

Each of the Commands has control over training and education establishments. For the most part, the mandate for these establishments is to prepare the soldiers, sailors, and air force personnel for combat operations. In view of the strategic situation, there is scope to review those mandates to include enhanced peacekeeping training objectives.

RECOMMENDATION 30. *It is recommended that the Commands review the mandates given to their staff colleges, warfare schools and similar institutions, with a view to enhancing the peacekeeping training objectives of those institutions.*

Land Force Area Headquarters

Many of the recommendations made for the Commands are also applicable to the area headquarters. However, the area headquarters have the main task of generating the personnel and equipment needed for the training of army deploying units.

RECOMMENDATION 31. *It is recommended that all Land Force Area Headquarters assume full responsibility for training and screening all augmentees so that they arrive at a deploying unit at the same level of general-purpose combat capability (battle task standard) as the personnel of the deploying unit.*

RECOMMENDATION 32. *It is recommended that Land Force Area Headquarters be the principal interface with non-military organizations, and be the channel for providing training assistance from those organizations (e.g., civilian police, Red Cross, Canadian peacekeeping partners, etc.).*

Headquarters Brigade Groups, and Similar Sea and Air Entities

Brigade headquarters is the most critical command level to help identify the training needs of regular units, and ensure that they can access the training and information they need for both regular training schedules and pre-deployment training. Brigade also plays the key role in ensuring that individuals they deploy are provided with sufficient and appropriate pre-deployment training.

There is an obvious link between the analysis of the mission and the training requirements. The brigade commander plays a key role in, first, determining those requirements, and second, confirming that the standards have been met before declaring the deploying unit to be operationally ready for its mission.

RECOMMENDATION 33. *It is recommended, particularly for a first-time deployment, that the brigade commander be the reconnaissance team leader, that reconnaissance take place before pre-deployment training commences, that the brigade commander assist the unit commander in the mission analysis, in prioritizing training requirements based on that analysis, and in conceptualizing, resourcing, and conducting unit exercises that will confirm that the requirements have been met.*

Battalions, Regiments, Air Squadrons, and Other Similar Size Units

While senior commands are key to facilitating peacekeeping training, the commander of any unit deploying is responsible for ensuring that the unit is adequately prepared. The unit commander's first obligation is to ensure that regular training schedules maintain the unit up to a satisfactory state of readiness. Then, when tasked with deploying, the unit commander must ensure that those peacekeeping skills are renewed and tailored for the particular mission.

Unit commanders also have primary responsibility for ensuring that individuals seconded from their unit for peacekeeping deployment are provided with sufficient and appropriate peacekeeping training.

No two units will begin the training preparation period at the same level, and all will require a "protected" period where they are free from other tasking and can concentrate on preparatory pre-deployment training.

The pre-deployment training, including the test evaluation, is highly dependent on four critical pieces of information: the mission; the command and control arrangements; the concept of operations; and the rules of engagement. Regardless of the political influences at play, this series of information must be given early to the deploying unit. If they are changed, an appropriate training period must be allowed for any readjustments to training.

RECOMMENDATION 34. *It is recommended that unit pre-deployment training time period be evaluated to ensure adequate generic peacekeeping training on subjects such as the law of armed conflict, negotiation procedures, low-level conflict resolution, and stress management, as well as mission-specific training on subjects such as concept of operations, rules of engagement, standing operating procedures, knowledge of theatre environment, and cultural awareness.*

The complexity of some peacekeeping training, particularly mission-specific training, requires subject matter experts that are readily available for pre-deployment training. Units that have recently completed a tour of duty in the same theatre are a valuable source of expertise. It appears logical that their training contribution be institutionalized for rotational deployments, so that units that have completed a tour are formally tasked, prior to their deployment, to be prepared to assist a follow-on unit (e.g., a unit proceeding on Rotation 2 is tasked to assist the unit on Rotation 4, etc.). Consideration should also be given to tasking a unit in-theatre to assist in the in-theatre training of a follow-on unit, either by leaving a rear cadre or by returning a training cadre. Full direct liaison regarding training should be authorized between in-theatre units and deploying units.

RECOMMENDATION 35. *It is recommended that units warned for deployment be fully supported by subject matter experts. These experts could come from the Land Force Command centres (e.g., the Peace Support Training Centre), be provided by the areas, and, as a matter of practice, come from units that have recently completed a tour of duty in the same peacekeeping mission.*

RECOMMENDATION 36. *It is recommended that much more effort be made by areas, brigades, and units to integrate non-military aspects of the UN mission (e.g., NGOs, UN agencies, CIVPOL) into the pre-deployment training, thereby making the peacekeeping partnership a true partnership.*

Canadian Forces Command and Staff College, and Canadian Land Force Command and Staff College

On their own initiative, these colleges have included some peacekeeping subjects in their curriculum. They are caught in the predicament of trying to increase this small percentage, while maintaining mandated training objectives in a finite training time. If that training time cannot be increased to accommodate peacekeeping subjects, there is a real requirement for senior-level prioritization. The Officer Professional Development Council should formalize training objectives for the Canadian Forces Command and Staff College, and LFC Headquarters should review its direction to the Canadian Forces Land Command and Staff College. Further, the colleges should consider whether they can modify existing training objectives, so that they are also placed in a peacekeeping scenario (e.g., ethics).

RECOMMENDATION 37. *It is recommended that the staff colleges increase their peacekeeping content by modifying their curriculum to include more non-traditional military training for peacekeeping, and to teach selected other subjects in a peacekeeping context. In addition, the colleges should include training with the other peacekeeping partners (CIVPOL, NGOs, UN agencies).*

Army Lessons Learned Centre and Proposed Peace Support Training Centre

One of the enduring observations from all interviews with units and individuals deployed on peacekeeping, was the fact that they felt they were in an information and peacekeeping training vacuum. The common theme was that they continually had to pull for information and training, as opposed to having too much pushed in their direction.

One part of the solution would be to have a CF responsibility centre that would be tasked with bringing together and updating all relevant information, intelligence, and lessons learned. They would then provide this in a user-friendly format to individuals, operational units, and training establishments.

The other part of the solution would be to have a training unit that has collected, created, or knows where to find a broad range of training modules, resources, subject matter experts, etc., all of which would be offered to individuals and units. Acting upon guidance from such a training unit, individuals and unit commanders could then identify what they required from what was being "pushed" at them. An ancillary role would be to provide the same resource to those training establishments that have peacekeeping modules.

As mentioned in Chapter 2, the army has a lessons learned centre, which, while it appears to have the right approach to its mandate, is, in fact, under-resourced and relatively unknown within the army and even more so in the other two services. It has recently been moved to Kingston and will be co-located with a complementary peace support training centre to be created in 1996. It should also be noted that there is an army simulation centre in Kingston that has potential to provide cost-effective peacekeeping training for field commanders, by allowing them to simulate field training in peacekeeping situations. Ostensibly, these co-located units will be able to meet many of the needs mentioned above, albeit predominantly for the army, with only tenuous links to the other services.

RECOMMENDATION 38. *It is recommended that the Army Lessons Learned Centre be provided with sufficient resources to collate and update relevant mission area(s) information and intelligence and provide this in a user-friendly format to individuals and units for pre-mission training. In addition, the lessons learned analysis output of the centre should be regularly transmitted to all training centres for inclusion into or correction of existing training.*

RECOMMENDATION 39. *It is recommended that the proposed Peace Support Training Centre be created as soon as possible, and that it be provided with sufficient resources to collect, create, or identify where to find a broad range of training modules, resources, subject matter experts, etc., and that all these be offered to individuals and units for pre-mission training. An ancillary role would be to offer these same resources to other training establishments.*

The study team fully supports the establishment of the Army Lessons Learned Centre and the proposed Peace Support Training Centre, but is concerned that these centres (operating under the command arrangements as the team understands them) will not satisfy national and tri-service requirements.

The study team understands that a proposal exists to create a tri-service joint task force headquarters based in Kingston. This proposal, coupled with the co-location in Kingston of the Army Lessons Learned Centre, the Peace Support Training Centre, the Army Simulation Centre and the Canadian Land Force Command and Staff College, provides an excellent opportunity to tie in under one commander most aspects of peacekeeping: doctrine, training, education, training support, training simulation, lessons learned, and real joint operations and operational planning.

RECOMMENDATION 40. *It is recommended that a review be undertaken to determine the feasibility of amalgamating the responsibilities of the JTF Headquarters, the Army Lessons Learned Centre, the Peace Support Training Centre, the Army Simulation Centre, and perhaps the Canadian Land Force Command and Staff College under one commander who would have accountability to Land Force Command for army matters, and to NDHQ for tri-service matters.*

Royal Military College

Training received at RMC is critical, as this is part of the initial or "basic" training for these entry-level officer cadets. In particular, the attitudes inculcated at this stage of training set an enduring theme or approach.

Quite aside from the influence upon RMC cadets themselves, an undeniable ripple effect is felt. Although RMC trains only about 25 per cent of all Canadian military officers, they ultimately fill a preponderance of senior command positions, and are a key influence in setting the philosophy and direction of the Canadian military. The training and attitudes delivered at RMC are, therefore, doubly critical.

Because of its unique Canadian military/academic character, RMC, more than any other university in Canada, should take a lead in creating peacekeeping credit courses that could be merged into most degree programs. RMC should also seriously consider creating an undergraduate degree in peacekeeping studies that would be interdisciplinary (history, political science, geography), in order to combine the military aspects with the myriad of related components of present-day UN peacekeeping.

RECOMMENDATION 41. *It is recommended that Royal Military College create peacekeeping academic credit courses, and that it seriously consider creating an undergraduate degree in peacekeeping studies.*

The military wing at RMC, as discussed in Chapter 2, does provide some peacekeeping-related training. At a minimum, the teaching of the law of armed conflict requires more than the present total of two hours over the four years that cadets spend at RMC. In addition to increasing the time devoted to teaching the law of armed conflict, the military wing should look at the new materials developed by the ICRC and other sources, in order to develop teaching modules that are grounded more in the operational application of humanitarian law than in a largely legalistic approach.

Related issues such as ethics, prejudice, etc., are important adjuncts to the law of war and good preparation for peacekeeping. The military wing of RMC should continue to focus on providing officer cadets with practical skills and tools to deal with ethics and humanitarian law in conflict situations. Practical examples of theory should be drawn not only from general ethical experiments or non-Canadian incidents or historical events (e.g., My Lai, World War I), but also from immediate operational examples, including peacekeeping missions, and, more specifically, from Canadian experiences such as the Oka crisis, and the Somalia peacekeeping mission.

RECOMMENDATION 42. *It is recommended that RMC cadets receive a minimum of two hours per year solely on the law of armed conflict. This, along with additional training in ethics and dealing with prejudice, should focus on their operational applications, rather than legal or theoretical overviews.*

Canadian Forces Recruit School and Other Basic Training Establishments

The narrow mandates for these schools does not allow much scope for increasing any type of peacekeeping training, with the exception of training on the law of armed conflict.

RECOMMENDATION 43. *It is recommended that all basic training establishments enhance their training on the law of armed conflict.*

Battle Schools

The army battle schools happen to be located in each of the army areas and are a useful general-purpose training resource for those areas. The study team believes that because battle schools have a major role in the follow-on training of army recruits, scope exists for them to conduct generic peacekeeping courses for others. In particular, the schools could teach individuals who are warned for UN duty (observers, etc.) and those augmentees (primarily Reserves) to warned units.

RECOMMENDATION 44. *It is recommended that the area battle schools be formally tasked to conduct peacekeeping training based upon a Land Force Command curriculum. These schools should also be given a mandate to assist other Commands that have peacekeeping tasks (e.g., helicopter squadrons), based on a curriculum developed by those Commands.*

Reserves

The "Total Force Concept" envisages an increased reliance upon Reserves to augment Regular Forces deploying on peacekeeping missions. Quite clearly, the high personnel turnover of Reserve units and the unevenness of individual and Reserve unit training has made their suitability problematic. This is not to denigrate the quality and commitment of many Reserve personnel who have gone on peacekeeping missions, but they would be the first to admit that many of their fellow Reserve members were not prepared to fulfil their peacekeeping roles.

This is, of course, part of a larger debate on the future role of Reserves. However, if the position is that Reserves will continue to provide a pool of personnel for peacekeeping, then they must be provided with substantially increased peacekeeping training, both the traditional military training received by the Regular Force,[18] and non-traditional military training for peacekeeping.

There are probably two general levels of training for most Reserves: steady state and enhanced. The steady state level must recognize the limitations of part-time service and develop basic standards. The enhanced level, on the other hand, must recognize that Reserve individuals will be placed in exactly the same operational situations as their Regular Force counterparts, and will require the same skills.

RECOMMENDATION 45. *It is recommended that the Reserves, in particular the Militia, review their training objectives with a view to including generic peacekeeping training. In addition, peacekeeping training standards need to be developed to support those objectives.*

73 Recommendations on Training Requirements

RECOMMENDATION 46. *It is recommended that Land Force areas assume full responsibility for the enhanced training of Militia augmentees, so that those individuals arrive at a deploying unit at the beginning of the pre-deployment period at the same level of general-purpose combat training and generic peacekeeping training as the unit personnel.*

Office of the Judge Advocate General

Training in the law of armed conflict is of critcal importance to effective peacekeeping: it cannot continue to be provided in an *ad hoc* manner. A clear responsibility centre must ensure that sufficient and effective training is conducted throughout the Canadian Forces.

Many of the subject matter experts on the law of war will come from JAG, since it forms the largest coherent CF grouping of such subject matter experts. Most others, particularly operators, would be drawn from a wide range of CF units. It appears logical, therefore, that JAG be tasked as the CF office of primary interest or responsibility centre.

Thus, JAG must receive sufficient resources, particularly personnel resources, to carry out both a large amount of writing, researching, preparation, and training of its own personnel, as well as the resources needed to fulfil its role of overseeing operators trained as subject matter experts, and ensuring their delivery of effective training. It might be useful to create a specific JAG teaching/research unit, and locate this unit close to the soon-to-be-created Peace Support Training Centre.

RECOMMENDATION 47. *It is recommended that there be a Chief of the Defence Staff directive to set out CF doctrine on the law of armed conflict, to emphasize the importance of training in the law of armed conflict, and to identify the Office of the Judge Advocate General as the responsibility centre for training on the law of armed conflict.*

The natural predilection of JAG will be towards the legalities of the law of armed conflict; JAG must endeavour to ensure that the operational ramifications of the law of armed conflict comprise at least half of the training for operators. This ratio should be even greater when training line personnel, increasing as one moves down the rank and command responsibility level. In addition, JAG should evaluate existing training, and follow up with recommendations on appropriate training content and location. In doing so, JAG should identify which individuals need which level of training. Critical in the navy would be ship commanders and boarding parties, in the air force, pilots and targeters. However, in the army, privates, corporals, and sergeants are often the first line of individuals applying the laws of war.

RECOMMENDATION 48. *It is recommended that all existing and future law of armed conflict training be primarily focused on integrating it into an operational context, and that operational military such as infantry officers and senior NCMs be trained to deliver much of that training.*

RECOMMENDATION 49. *It is recommended that the behavioural aspect of the law of armed conflict be recognized so as to make its teaching an integral part of basic training for all CF personnel, and that there be regular refresher training on this "basic attitudinal training" on the law of armed conflict.*

RECOMMENDATION 50. *It is recommended that the Office of the Judge Advocate General be tasked to identify the type and level of special training required for those exercising command functions that are reasonably likely to involve them in dealing with the interpretation and application of the law of armed conflict. In peacekeeping, such individuals invariably include corporals and sergeants.*

RECOMMENDATION 51. *It is recommended that the Office of the Judge Advocate General be tasked with overseeing the creation of mission-specific law of armed conflict training that would consist of short refresher courses with particular adaptations or guidance on its application for a particular peacekeeping mission.*

As noted in Chapter 2, CF training resources for the law of armed conflict are few and some resources are outdated. For example, the Law of Armed Conflict Manual was an excellent draft in 1986, but remains in draft form, and portions have now become quite dated. The New Zealand military recognized the worth of this draft, and used it as the basis for their own such manual. As Canada is part of a group of countries[19] that are reworking their manuals, JAG should identify sufficient staff resources to rework the 1986 draft manual and publish it in pamphlet form. Subsequently, JAG should rework this type of publication and related materials at least every 10 years.

RECOMMENDATION 52. *It is recommended that the Office of the Judge Advocate General, in conjunction with various CF training establishments, update or create training curriculum and resources. The Office of the Judge Advocate General should also be encouraged to complete the rewrite and publishing of its 1986 draft Law of Armed Conflict Manual.*

The International Committee of the Red Cross is the world expert on the law of armed conflict, and works hard to encourage militaries to train on the law of armed conflict. When resources allow, the ICRC conducts training itself, but it prefers to train the trainers. It has recently developed extensive new law of armed conflict training material, including a training exercise, at the battalion level. ICRC training is placed very much in an operational military context as opposed to a legal approach, and training is provided by operational military (combat arms) rather than legal officers. A key ICRC objective is to set out the link between the law of armed conflict and sound tactical decisions.

RECOMMENDATION 53. *It is recommended that the Office of the Judge Advocate General draw upon the expertise and involvement of the International Committee of the Red Cross in the design and delivery of law of armed conflict training.*

OTHER SOURCES OF TRAINING PRACTICE AND GUIDANCE

Pearson Peacekeeping Centre

In 1994, the Canadian government provided funding for the creation of an independent and freestanding Pearson Peacekeeping Centre (PPC). It has a specific mandate to carry out joint peacekeeping training for military and civilian personnel. Inter-operability or inter-agency cooperation in peacekeeping operations is increasingly a critical factor, and the military-civilian divide is, of course, often the most difficult to bridge. PPC appears to be the first training establishment in the world to formally marry the two in its training goals and audience.

To date, PPC has offered a large number of two-week courses on issues such as peacekeeping partners, mediation, refugees, naval peacekeeping, legal aspects, and a 10-week peacekeeping management, command, and staff course. They are also designing a course on human rights in peacekeeping. Despite the inevitable start-up difficulties and its isolated location,

PPC has attracted international attention and raised expectations, largely because of its composite civilian/military character.[20] The PPC will logically continue to focus on this unique and rapidly growing civilian/military market niche.

A large number of Canadian military, and foreign military under the Military Training Assistance Program, have and should continue to attend PPC courses, for a number of reasons. First, PPC is providing training to CF members that, for the most part, does not yet exist within the CF training constellation. Secondly, the CF can draw expertise on designing, delivering, and regularly updating its own in-house peacekeeping training from PPC courses. Thirdly, even with future comprehensive CF peacekeeping training, PPC will continue to offer that important joint military-civilian training environment. Fourthly, as a unique Canadian training establishment designed to serve the international community and disseminate Canadian peacekeeping expertise, the participation of the Canadian military is a critical contribution to the character and content of PPC training.

RECOMMENDATION 54. *It is recommended that the Canadian Forces continue to send its members to all of the Pearson Peacekeeping Centre's courses in order to: train CF members; gain additional expertise to develop CF peacekeeping training; train in the centre's unique civilian-military training environment; and contribute to the civilian-military character and content of the centre's training.*

Such CF involvement in PPC training should be distinct from peacekeeping training for Canadian Forces *per se*. It was never intended that the PPC fulfil the requirement for comprehensive Canadian Forces in-house training.

It was beyond the scope of this study to examine other independent peacekeeping institutions that are broadly similar to the PPC but have largely civilian audiences, e.g., the European University Centre for Peace Studies at Stadtschlaining, Austria. It would be beneficial for the Canadian Forces to look at the usefulness of CF personnel visiting or attending some of those institutions' training courses.

Canadian Civilian Police

United Nations civilian police make up an increasingly large component of UN peacekeeping operations. The general increase in the use of CIVPOL as peacekeepers stems from the immediate applicability to peacekeeping tasks of the domestic experience and training of many national police. Canadian CIVPOL[21] have proven to be particularly adept at UN peacekeeping.

For a number of reasons, including the fact that most Canadian police officers have full-time policing jobs that do not allow them the luxury of lengthy pre-mission training, Canadian CIVPOL "units" receive about one week of mission-specific training prior to deployment, and another three days of induction training upon arrival in the field. Individuals receive less, although there is a determined effort to provide them with three days of pre-deployment training. Thus this training is rushed and, in itself, does not provide substantial lessons for Canadian military peacekeeping training.

However, the key skill set that makes Canadian CIVPOL so useful in UN peacekeeping operations is their attitude and approach to policing in general. This attitude and approach, as developed through ongoing regular police training, should be of special interest to Canadian military designing peacekeeping training. It includes issues such as minimum use of force; self-defence as opposed to any attack mode; using verbal and other behavioural skills to avoid violence; working for, through and with the public; and working within extensive legal limitations combined with a regular and often quite extensive public overview. With the RCMP for example, this indoctrination training starts at cadet training in Regina, and is carried on throughout subsequent formal and on-job training.

Canadian police are taught a variety of skills to contain or limit violence, as well as an awareness of how to gradually move through the stages of escalating force as required. A critical component of this training is independent analysis and decision-making by officers to determine the minimal level of force ("lowest level of resolution") necessary for both self-protection and a satisfactory resolution of the incident at hand. They also adhere to a client-centred approach, popularly known as "community policing." In the peacekeeping context, this translates into a firmly held attitude that the parties to the conflict are not the enemy, and that CIVPOL response to conflict situations should involve the minimum force necessary.

Inasmuch as low-level conflict, either in peacekeeping or in aid to the civil power, makes up most of the Canadian Forces present operational activity, it makes sense that the CF consider training to deal with a broader range and more calibrated use of force. Canadian military are traditionally trained on the higher level use of force, but are not as well versed in the broader police range use of force. In addition, the CF should aim to increase the ability of Canadian military to move from combat mode to a "minimum use of force" mode.

RECOMMENDATION 55. *It is recommended that the Canadian Forces look at the training provided to Canadian police in the determination of the minimum use of force necessary, the broad range of use of force options, and how to gradually escalate and de-escalate through this range.*

Aside from these attitudes and operating skills that are incorporated into most Canadian police training and daily experiences, Canadian police undergo specialized training at the detachment level and centrally. For example, the Canadian Police College conducts about 25 different courses. All training is specific to domestic police functions, but some courses, such as those dealing with mediation and negotiating,[22] have a definite peacekeeping use and may well lend themselves to military participation and/or as a model to be modified and adapted to meet military peacekeeping needs.

"Officer survival" training is a regular part of domestic police training. The purer form of officer survival has much to do with personal tactical approaches to conflict situations. Training includes situation-specific applications in the use varying degrees of force and self-defence techniques. The broader range of officer survival training consists of mediation and negotiating skills, which help officers "survive" by talking themselves out of trouble and/or resolving a dispute. Aside from this training for domestic policing, most Canadian CIVPOL receive mission-specific refresher survival training for a particular peacekeeping cultural/political context.

RECOMMENDATION 56. *It is recommended that the Canadian Forces look at the potential usefulness of some Canadian police training in areas such as mediation and negotiation, and officer survival training, including pre-deployment mission-specific officer survival training provided to Canadian CIVPOL.*

Auditor General

The Office of the Auditor General regularly audits defence management, and has a dedicated team that for several years has been looking at a variety of topics that touch directly on peacekeeping issues and indirectly on training for peacekeeping. The 1994 Report of the Auditor General stated that "senior force development planners found that the existing policy development system did not provide sufficient guidance on the types of conflict the Canadian Forces should be prepared for. Hence they had difficulty in defining the forces that would meet Canada's needs."[23] This lack of sufficient guidance on which roles the Canadian Forces should prepare for may partly explain why those responsible for CF training have continued to see peacekeeping as largely a peripheral CF role, and thus a peripheral training requirement.

The Auditor General team tasked with National Defence audits is increasing its focus on issues such as training, and its current audit of peacekeeping has "training for peacekeeping" as a specific sub-audit. *Inter alia* the team will be assessing the degree to which systems are in place to ensure adequate, appropriate, and consistent training. It is far too early to anticipate the eventual conclusions of the audit, and the team has stressed that it approaches its task from a cost-effectiveness perspective, rather than as an attempt to assess the strategic and political benefits of any particular activity.

RECOMMENDATION 57. *It is recommended that the Canadian Forces monitor the Auditor General's current sub-audit on training for peacekeeping for lessons and ideas.*

Other Countries' Military Training for Peacekeeping

This study was not able to investigate the training that other countries provide to their military peacekeepers. Information the team did gather was therefore both partial and anecdotal. Some comparative reports were consulted, but from the comments on Canadian training for peacekeepers, it was obvious that these were based on a review of documentation, rather than actual training being conducted. As a result, these reports, in part, commented on what countries hoped to do, rather than what they were actually doing.

A short but interesting overview of Scandinavian peacekeeping training by Langille/Simpson[24] does state, however, that "the Scandinavian training programmes focus on the skills and requirements of UN service. Aside from the general peacekeeping training programme for all ranks, they conduct special officer courses and mission training courses. Within their training centres, there is a consensus that peacekeeping training is clearly distinct from the ordinary military training which a soldier receives."²⁵ The overview discusses cooperative training program conducted by Denmark, Sweden, Finland, and Norway, and concludes that it "is widely recognized as the most advanced in the world."²⁶

Training is also being carried out by countries such as the Netherlands, Austria, the U.K., Australia, New Zealand, and the U.S.A. The U.S.A., with its substantial resources, is moving very quickly to develop extensive peacekeeping doctrine and training. Recent developments saw that countries such as Poland and the Czech Republic are creating dedicated peacekeeping training schools for their military. However, it must be repeated that a list on paper of countries carrying out peacekeeping training, along with their training curriculum, does not necessarily reflect the quality and usefulness of such training.

Nevertheless, the study team's peripheral investigation into peacekeeping training in other countries made it quite clear that Canada must assume that its specialized training for peacekeeping places it in the middle of the pack. In particular, various European nations have been carrying out peacekeeping-specific training for quite some time, and recently have been revamping and increasing the type and degree of such training. It will be a useful exercise for the CF to assess other militaries' peacekeeping training for ideas on Canadian training.

As referred to in Chapter 2 and discussed further below, the UN, with Canadian NDHQ involvement, is involved in a survey and assessment of peacekeeping training throughout the world. This study will be of use to the Canadian Forces, signalling, in particular, which training programs Canada should investigate in depth. There are certainly training programs and curriculum that exist in other countries that Canada would be well advised to adapt to Canadian needs.

RECOMMENDATION 58. *It is recommended that Canada follow closely the UN Department of Peacekeeping Operations study on peacekeeping training by all member states, and that the Canadian Forces use that and other indicators as guidance on other militaries' peacekeeping training, in order to assess them in depth and with a view to adapting those programs and materials to Canadian peacekeeping training needs.*

United Nations

Multinational and multidimensional UN peacekeeping operations contain a multiplicity of UN and national standards and operating procedures. It is absolutely essential that common training and common standards be established for UN peacekeepers. However, it has never been intended that the UN train national militaries for peacekeeping, certainly not militaries such as Canada's.

The UN has consistently maintained that "in view of the large numbers involved, the training of personnel provided by Member States will remain primarily the responsibility of Governments....based on common standards and a common curriculum."[27] A recent DND/Foreign Affairs study agreed fully with this position, and stated that "DPKO should take the lead in the development of training standards and guidance to troop contributing nations, as well as an ability to ensure that Member States are adhering to training standards consistent with UN requirements."[28]

Historically, a resource-starved UN DPKO has found it difficult to meet this challenge of facilitating troop contributing nations in their training for peacekeeping. However, the DPKO Training Unit has worked hard and continues to harness its limited resources with a view to carrying out such a role.[29] It is developing an impressive array of manuals and training curriculum in conjunction with other parts of DPKO and various troop contributing nations. The Training Unit's publication plan has 27 titles that already exist and are being improved upon, or that are being drafted or about to be drafted.[30] In addition, DPKO held a major regional training course in September 1994 for senior-level officers, and three more regional courses are being planned. DPKO have identified command post exercises as useful training vehicles, and are moving to develop such exercises. In all of this, DPKO and its training unit appear to understand the importance of non-traditional military subjects.

What this means is that the UN itself cannot, at present, contribute substantially to the evolution of Canadian Forces peacekeeping training. Along with several other troop contributing nations, Canada must play a greater role in devising new training for itself and sharing it with the UN.

Over time, as UN training materials and standards become better established, there will be an increasing need for a UN cadre of trainers who, while continuing to devise training materials, will be training the trainers of many troop contributing nations to carry out all or most of their own in-country training. In fact, the DND/Foreign Affairs study recognized the importance of this and referred to DPKO's "UNTAT [training assistance teams] System, which, in addition to current duties, could try to identify training needs in Member States and develop ways for redressing any training gaps."[31]

Canada is a key participant in the UNTAT program, and NDHQ has identified a responsibility centre in NDHQ which is actively involved in the current UNTAT exercise of determining what kind of training and training resources exist among UN countries. The goal is to draw on such national expertise to create or enhance UN training materials, and subsequently training assistance to various troop contributing nations.

RECOMMENDATION 59. *It is recommended that as the Canadian Forces develop new non-traditional military training for peacekeeping, they share course packages, training materials, etc. with the UN and other troop contributing nations.*

RECOMMENDATION 60. *It is recommended that Canada continue to play a major role in assisting UN DPKO, through such vehicles as training assistance teams, to develop and deliver training standards, materials, and assistance to a variety of other troop contributing nations.*

CHAPTER FOUR

Conclusion

More than 15[1] Canadian service personnel have died in the service of peace since 1990. The Canadian Forces have a world-renowned record as peacekeepers, due in large part to their professionalism and their general-purpose combat training. Recent events in Rwanda, Bosnia Herzegovina, and Croatia have demonstrated both the wisdom of that emphasis as well as the changing nature of peacekeeping missions.

Traditional roles found in Cyprus-type peacekeeping have been augmented by a complex matrix of new and unfamiliar tasks, which demand training and skills beyond general-purpose combat capability. In addition, events occurring during the Canadian Forces deployment to Somalia have called into question the effectiveness of some aspects of Canadian preparation for peacekeeping missions.

The study team found a dichotomy. On the one hand, there were many separate and unconnected examples of Canadian Forces organizations and individuals who understood the changing peacekeeping environment and peacekeeping requirements, and were taking some initiatives to meet those needs. On the other hand, the study team could not ascertain a national, formalized, coherent, integrated peacekeeping policy and training program that did likewise.

The study team has concluded that within the Canadian Forces, the conviction that a well trained combat capable soldier is *all* that is required for a good peacekeeper is changing or at least being modified. However, the bureaucracy has not yet caught up with the changing philosophy. A methodical process is needed that involves a review of government peacekeeping policy and requirements; the formulation of a national level Canadian Forces peacekeeping training policy; the writing of peacekeeping doctrine to support both policies; the setting of peacekeeping standards and priorities; and the allocation of responsibilities and sufficient resources.

Instigating any one of these steps in isolation, while commendable, risks an overall lack of priority and continuing indifference by those not directly involved. The single most important step is for the political authorities and senior military to formally acknowledge that "business as usual" cannot continue, and to initiate a systematic policy change.

Until recently, the training system used by the Canadian Forces has provided effective peacekeepers where traditional peacekeeping was sufficient. However, second-generation peacekeeping for increasingly complex emergencies requires the Canadian military training system to adjust to new circumstances and new requirements. This study has attempted to outline training shortfalls and to recommend possible solutions. The objective of the study team has been in keeping with the objectives of the Canadian Forces: how to train the best Canadian peacekeepers possible.

Perhaps we can be guided by the words of UN Secretary-General Boutrous Boutrous-Ghali: "Reform is a continuing process, and improvement can have no limit. We must be guided not by precedents alone, however wise these may be, but by the needs of the future and by the shape and content that we wish to give it."[2]

Notes

CHAPTER ONE — INTRODUCTION: RATIONALE AND POLICY

1 Some of the comments from those who had requested such training were made directly to study team members. More often, we received comments from individuals referring to requests they had received and been unable to meet, or requests they knew others had made for assistance in peacekeeping training.
2 Peacekeeping has become a variable term. This study uses the term in its broadest sense to include peace-making, peace-enforcement, peace-building.
3 Blechman and Vaccaro, *Training for Peacekeeping: The United Nations' Role*, Report No. 12 (The Henry L. Stimson Centre, July 1994), p. 15. The authors note the superficiality of their analysis, which *inter alia*, compared casualty rates in UNPROFOR for Belgian peacekeepers with no specialized peacekeeping training and Danish peacekeepers, who receive substantial specialized training and have had proportionally far fewer casualties.
4 UN term for civilian police.
5 Non-governmental organizations; International Committee of the Red Cross.
6 UN High Commissioner for Refugees; World Food Program, UN Development Program.
7 Thomas Weiss and Kurt Campbell, "Military Humanitarianism", in *Survival* XXXIII/5 (September/October 1991), p. 457.
8 Frédéric de Mulinen, *The Law of War and the Armed Forces*, Series Ius in Bello, No. 1 (Geneva: Henry Dunant Institute, 1992).
9 Frits Kalshoven, *Constraints on the Waging of War* (Geneva: International Committee of the Red Cross, 1987), p. 61.
10 House of Commons, Senate Subcommittee on Security and National Defence, *Meeting New Challenges: Canada's Response to a New*

Generation of Peacekeeping, Report of the Standing Senate Committee on Foreign Affairs (February 1993).
11 House of Commons, Standing Committee on National Defence and Veterans Affairs, *The Dilemmas of a Committed Peacekeeper: Canada and the Renewal of Peacekeeping* (June 1993).
12 Parliament of Canada, Special Joint Committee on Canada's Defence Policy, *Security in a Changing World* (25 October 1994).
13 *Meeting New Challenges*, p. 10.
14 *Meeting New Challenges*, p. 11.
15 *Meeting New Challenges*, p. 11.
16 *The Dilemmas of a Committed Peacekeeper*, p. 25.
17 *Security in a Changing World*, p. 22.
18 Department of National Defence, *White Paper* (1994), p. 34 [hereafter, *White Paper* (1994)].
19 *White Paper* (1994), p. 34.
20 Government of Canada, "Towards a Rapid Reaction Capability for the United Nations", Report of the Government of Canada (September 1995).
21 Final Report on the in-depth evaluation of peacekeeping operations: start-up phase, E/AC.51/1995/2, paragraph 89. It is also effective UN doctrine: see General Assembly Resolution 48/42, paragraph 45: "the training of peace-keeping personnel is primarily the responsibility of Member States."
22 A/50/60, paragraph 45.

CHAPTER TWO — TRAINING IN THE CANADIAN MILITARY

1 Canadian Forces Individual Training System (CFITS), Introduction (A-P-000-001/PT-000), 31 July 1989, p. 1-1-1.
2 CFITS, Introduction, p. 1-1-1.
3 Management of Training (A-P9-000-002/PT-000), volume 2, 29 March 1991, p. 1-5.
4 Canadian Forces Administrative Order 9-47, 29 December 1987, p. 7.
5 Final Report on NDHQ Program Evaluation E2/90: Peacekeeping, p. xvi, paragraph 21 [hereafter, Final Report].
6 Final Report, p. xvii, paragraph 23.
7 Final Report, p. 192, paragraph 4.146.
8 Final Report, p. 199, paragraph 4.167.
9 Final Report, p. 252, paragraph 4.330.
10 Final Report, p. 253, paragraph 4.335.

11 Final Report, p. 255, paragraph 4.339.
12 J3 PK 155 252029Z (March 1993).
13 Set out in J3 PK 155 252029Z (March 1993). See also UN Training Guidelines for National or Regional Training Programs, undated UN document produced by UN DPKO Training Unit.
14 *Meeting New Challenges: Canada's Response to a New Generation of Peacekeeping.*
15 *The Dilemmas of a Committed Peacekeeper: Canada and the Renewal of Peacekeeping.*
16 4500-1 (DCDS) 29 December 1993, Training Requirements for Peacekeeping Operations, p. A-4, paragraph 3. See Appendix I to this study.
17 4500-1 (DCDS), p. A-4, paragraph 4.
18 4500-1 (DCDS), p. A-6, paragraph 8.
19 4500-1 (DCDS), p. A-7, paragraph 10.
20 4500-1 (DCDS), p. 2, paragraph 4.
21 4500-1 (DCDS), p. 3, paragraph 5.
22 This lack of follow-up appears to have been fairly consistent. For example, a Vice Chief of Defence Staff directive, 3451-9 (VCDS) of December 11, 1992, directed that peacekeeping training issues be reviewed "with a view that they be formalized and that responsibilities be carefully delineated". See 4500-1 (DCDS), 29 December 1993, p. 1.
23 For a compilation of some of the past internal DND warnings and efforts to adapt training to peacekeeping needs, see P. Langille, *Consolidating Canadian Forces' Peacekeeping Training Efforts*, submission to the Special Joint Committee on Canada's Defence Policy, 2 August 1994.
24 The study team could not locate examples of national-level Canadian peacekeeping doctrine produced before 1994.
25 CFP(J)5(4), Joint Doctrine for Canadian Forces Joint and Combined Operations, p. 21-16, paragraph 2111.
26 CFP(J)5(4), pp. 21-24 to 21-26, paragraph 2118.
27 B-GL-301-033/FP-001, Peacekeeping Operations, p. 11-1-1, paragraph 4.
28 B-GL-301-033/FP-001, p. 11-3-1, paragraph 2.
29 B-GL-301-033/FP-001, p. 11-3-2, paragraph 4.
30 See B-GL-301-033/FP-001, p. 11-6-12, paragraph 2.
31 Visits and interviews by team members took place between May and September 1995, with most occurring in June and July 1995.
32 Interview J3 Training, NDHQ, 26 June 1995.
33 Interview J3 Training.
34 See 4500-1 (DCDS), 2 January 1995.

88 Notes for pages 30–48

35 The RCR commanded by LCol. M.S. Skidmore deployed to Croatia on 1 October 1994, for six months; the RCD commanded by LCol. W Brough deployed to Bosnia 1 November 1994, for six months. The brigade commander was BGen. N.B. Jefferies.
36 Canadian Land Force Command and Staff College course brief OUE/06/P, p. 43-44.
37 4500-1 (DCDS), 29 December 1993, p. A-11, paragraph k.
38 3120-55-5-1 (G3 Trg Dev 2), June 1995, p. 1.
39 3120-55-5-1 (G3 Trg Dev 2), June 1995, p. 1.
40 3120-55-5-1 (G3 Trg Dev 2), June 1995, p. 2.
41 The name given to the provision of service support for the Canadian Contingent of UNPROFOR. When tasked by LFCA, the RCR BS CO is responsible.
42 Variously titled the law of armed conflict, the law of war, international humanitarian law, etc., this body of international norms includes the Geneva Conventions (Law of Geneva), the Hague Conventions (Law of the Hague), and the more recent Law of New York. This study uses the term 'law of armed conflict' to refer to all of these.

CHAPTER THREE — RECOMMENDATIONS ON TRAINING REQUIREMENT

1 *White Paper* (1994), p. 13.
2 *White Paper* (1994), p. 14.
3 Discipline is central to any form of professionalism, as well as success in any endeavour, including military operations and peacekeeping operations. The study team considered this so self-evident that it was not necessary to deal with it for purposes of this study.
4 Various combat skills are at more of a premium in certain peacekeeping situations. Our research identified some traditional military skills, such as urban patrolling and convoy security, that would apparently benefit from greater priority to prepare Canadian peacekeepers better for particular missions. However, with the exception of the laws of war, our mandate and focus did not result in us examining the priority given to traditional military skills in this regard.
5 "Towards a Rapid Reaction Capability for the United Nations", p. 53.
6 UN term for civilian police.
7 Non-governmental organizations; International Commi tee of the Red Cross.
8 UN High Commissioner for Refugees; World Food Program, UN Development Program.

9 "Towards a Rapid Reaction Capability for the United Nations".
10 Possibly through the Canadian Forces Language School in Ottawa.
11 An interesting aspect of this was reported concerning low-level conflict resolution scenarios. Many mediation/negotiation techniques aim at achieving long delays or cooling off periods to allow time for partial healing. Because of the limited time available to conduct a mediation field exercise during the 90-day pre-deployment training period, however, some individuals were left with a mistaken impression of how long mediation/negotiation activities would actually take in the field; as a result, attempts work out solutions too quickly ran the risk of failure.
12 Several Canadian peacekeepers talked to the New Zealand contingent and were told that they receive specific training in negotiation and conflict management.
13 *White Paper* (1994), pp. 32-33.
14 Variously called the law of armed conflict, the law of war, international humanitarian law, etc., this body of international norms includes the Geneva Conventions (Law of Geneva), the Hague Conventions (Law of the Hague), and the more recent Law of New York. This study uses the term 'law of armed conflict' to refer to all of these.
15 Blechman and Vaccaro, "Training for Peacekeeping: The United Nations' Role", p. 4.
16 For example, current OPDP coverage of the law of armed conflict needs to be expanded.
17 Standard Minimum Rules for the Treatment of Prisoners; Procedures for the Effective Implementation of the Standard Minimum Rules for the Treatment of Prisoners; Body of Principles for the Protection of All Persons Under Any Form of Detention or Imprisonment; Code of Conduct for Law Enforcement Officials; Guidelines for the Effective Implementation of the Code of Conduct for Law Enforcement Officials; Basic Principles on the Use of Force and Firearms by Law Enforcement Officials; UN Criminal Justice Standards for Peace-keeping Police; Principles for the Effective Prevention and Investigation of Extra-Legal, Arbitrary and Summary Executions.
18 An example of the paler form of training received by reserves is the Militia Command Staff Course, which consists of self-study (Part I), district training and examinations (Part II), all leading to a three-week Militia Command and Staff Course in Kingston in the summer. This ostensibly compares with the very intensive six-month Canadian Land Forces Command and Staff College course.

19 Canada, New Zealand, Australia, the United Kingdom, and the United States.
20 The UN Special Committee on Peace-keeping Operations called specifically for "the establishment of peace-keeping training centres...for military and civilian personnel", A/49/136, p. 16.
21 Canadian UN CIVPOL are composed largely of RCMP members, but there are sizable additions from other Canadian police forces. While we do not intend to minimize the contribution of these other forces, the RCMP has been the lead agency in this area, and time constraints limited the study team to interviews with RCMP personnel.
22 It was suggested that a particularly strong three-day negotiator course was given by the Canadian Centre for Management Development. The Canadian Police College also has a domestic hostage negotiating course. It should be noted that the Pearson Peacekeeping Centre has developed a mediation module as well.
23 *Report of the Auditor General of Canada to the House of Commons, 1994*, volume 15, pp. 2413.
24 Peter Langille and Erika Simpson, *A 1994 Blueprint for a Canadian and Multinational Peacekeeping Training Centre at CFB Cornwallis*. See Annex M: "Overview of the Scandinavian Peacekeeping Training Programme and Training Centres", a Submission by the Province of Nova Scotia to the Special Joint Committee of the Senate and the House of Commons on Canada's Defence Policy, 1994, p. 21A:99-106.
25 Langille and Simpson, "Overview", p. 21A:105.
26 Langille and Simpson, "Overview", p. 21A:105.
27 *Improving the Capacity of the United Nations for Peace-keeping*, Report of the UN Secretary-General, A/48/403 (S/26450), 14 March 1994, paragraphs 23-24.
28 "Towards a Rapid Reaction Capability for the United Nations", p. 54.
29 For a fuller overview of UN peacekeeping training, see P. LaRose-Edwards, *United Nations Internal Impediments to Peace-keeping Rapid Reaction* (Department of Foreign Affairs, 2 April 1995), p. 54.
30 United Nations: Guidelines for Peace-keeping, Code of Conduct, Peace-keeping Training Guidelines, Infantry Units Field Manual, Peace-keeping Curriculum for Staff College, Military Observers Handbook, Military Observers Course, Civilian Police Course, Civilian Police Handbook, Training of Local Police, Staff Officers Course, Staff Officers Handbook, Peacekeeping Course Calendar, Peace-keeping Bibliography, Stress Management, Medical Units Manual, Peace-keeping Terminology

Notes for pages 81–84

Handbook, Communications Manual, Movement Manual, Junior Ranks Handbook, Peace-keeping Negotiation Handbook, Operational Support Manual, Engineer Support Manual, Logistics Officers Course, Military Police Course, Training Centre List.
31 "Towards a Rapid Reaction Capability for the United Nations", p. 54.

CHAPTER FOUR — CONCLUSION

1 Canadian Forces International Operations (May 1995), p. 1-8-4/5.
2 Boutrous Boutrous-Ghali, *An Agenda for Peace*, A/47/277, 17 June 1992, p. 48.

APPENDIX I

4500-1 (DCDS) 29 Dec 1993

Training Requirements for Peacekeeping Operations

This directive was issued by the Deputy Chief of the Defence Staff with the aim of improving peacekeeping training in line with the recommendations of several parliamentary committees and internal DND evaluations (see Chapter 2).

4500-1 (DCDS)

29 December 1993

Distribution List

TRAINING REQUIREMENTS FOR PEACEKEEPING OPERATIONS

References: A. 3451-9 (VCDS)
11 December 1992
B. 4500-1 (DCDS ISO)
10 January 1993
C. UN Trg Guidelines for National or Regional Trg Programmes 91-02208
8 February 1991
D. J3 Pk 155 252029Z
March 1993

1. The nature of peacekeeping is changing and training for peacekeeping remains both an issue of significant domestic and international interest, and one of critical operational importance to the Canadian Forces. The CF has enjoyed an excellent reputation both at home and abroad for a highly professional and significant contribution to peacekeeping operations for more than forty years.

2. Recently, the effectiveness of CF peacekeeping training has been questioned. The Standing Committee on National Defence and Veteran's Affairs (SCONDVA) report of June 93 and an internal Chief of

94   Appendix I

Review Services study both recommend changes to peacekeeping training in the CF. Reference A directed that peacekeeping training issues be reviewed "with a view that they be formalized and that responsibilities be carefully delineated". Reference B directed that J3 Peacekeeping (J3 Pk) coordinate the staff action that would summarize current peacekeeping related training, identify further training requirements and provide recommendations on improvements.

3. To rectify deficiencies and implement improvements this letter will formalize the guidance and direction for preparing and training for peacekeeping operations. Some general improvements to established training sequences and direction to develop new programmes are at Annex A. All personnel selected for deployment will receive the training outlined in Annex A, Appendices 1 and 2, as applicable to the mission and unit size. These syllabi are proposed as minimums only, as it is recognised that individual commanders will need to expand or modify the programmes to meet specific mission and unit needs. The instructions contained herein will be incorporated into an omnibus DCDS instruction on deployed operations to be released early in 1994.

4. Commands are expected to determine additional training requirements as dictated by the mission and the initial state of training of assigned personnel, and to identify where changes and improvements can be made. Since personnel are expected to arrive at the training sessions current in basic military skills, priority is to be given to mission specific information such as the UN policy, ROEs, the geo-political and military situation and cross-cultural awareness. In the case of formed units, attention must be focused on the development of collective operational skills and cohesion. Where personnel arrive lacking in basic military skills, additional training will be required.

5. Additional mission dependent skills, such as mine awareness training, convoy escort duties or cordon and search operations will be taught as required. Verification or instruction of small arms skills, physical fitness and first aid will also take place during the training programme. The training directive that is issued as a supplement to the NDHQ tasking order will outline any additional training requirements. Confirmation by unit commanders that their personnel are operationally ready to deploy will indicate they are trained to the standard set by their respective Command and approved by NDHQ.

Appendix I

6. Implementation of this directive is to begin upon receipt of this document. Command Headquarters are to ensure that this instruction is distributed as required. Implementation is to be reviewed periodically by J3 Pk who are to prepare an interim report by Aug 1, 1994 and a comprehensive report by Aug 1, 1995.

[signed]
L.E. Murray
Vice-Admiral
for Chief of the Defence Staff

Annexes:
Annex A Training for Peacekeeping Operations.

DISTRIBUTION LIST

Action	*Internal*	*Information*
External	Surg Gen	CDS
MARCOM/Comd	DGPCO	DM
LFC/Comd	DGPCOR	VCDS
AIRCOM/Comd	DGRET	CRS
CFCC/Comd	C Res & Cdts	ADM(Pol & Comm)
CFTS/Comd	JAG	ADM(Per)
NRHQ/Comd		ADM(Mat)
3 CSG/CO		ADM(Fin)
		CPCD
		DGPA
		DG Exec Sec
		D Hist

ANNEX A
TO 4500-1 (DCDS)
29 DECEMBER 1993

TRAINING FOR PEACEKEEPING OPERATIONS

AIM

1. The aim of this document is to provide direction on training for peacekeeping operations.

PEACEKEEPING ROLES AND TASKS

2. Peacekeeping has evolved from the traditional and recognized method of inserting noncombat observers between two or more disputive but agreeable parties. The United Nations has divided peacekeeping into different but overlapping sub-categories that require diverse approaches to training and implementation. The categories of peacekeeping to which elements of the Canadian Forces may be deployed are:

   a. *Preventive Diplomacy.* This refers to a range of diplomatic activity designed to identify areas of potential conflict and pre-empt the outbreak of fighting. These activities could include confidence building measures such as the monitoring of regional arms agreements, exchange of military missions, fact finding missions and the establishing of demilitarized zones;

   b. *Preventive Deployment.* As a follow on to preventive diplomacy, or in addition to it, the UN may establish a military presence between states upon the request of one or more parties to the dispute to act as a deterrence to conflict. The deployment of a Canadian company group organization to Macedonia in Jan-Feb '93 was the first example of employing a UN military force in a preventive deterrence role;

   c. *Peacemaking.* These are activities normally conducted after the commencement of conflict and are designed to bring the warring parties to agreement by peaceful means. Canada considers peacemaking to be principally diplomatic activity, but considers peace restoring or peace enforcement activities to belong in this category. Examples of such activities include the UN missions in Korea (1950-53), the Congo (1960-64) and Somalia (92-93);

d. *Peacekeeping*. This involves the deployment of a UN presence in the field, often consisting of military, police personnel and civilians, to observe and supervise a recognized truce or ceasefire. Unlike peace enforcement units, peacekeeping units are usually relatively lightly armed, possessing enough weaponry for self-defence only. Troops deployed in Cyprus (UNFICYP) met the definition of peacekeeping forces;

e. *Peace-enforcement*. In the event that cease-fires have been agreed to but not complied with and/or the UN has been called upon to restore and maintain order, the Secretary General may call for the deployment of peace-enforcement units. These units would be more heavily armed than peacekeeping units and would require extensive training and preparation. No UN missions have been deployed to date under this specific mandate;

f. *Peace-building*. These are usually post-conflict activities designed to identify and support structures which will tend to strengthen and solidify peace in order to avoid a relapse into conflict. They encompass a range of activities from monitoring of elections (eg, UNTAC) to the re-building of a country's infrastructure (UNMIH in Haiti). Most peacekeeping missions include some peace-building activities as part of their mandate;

g. *Observer Missions*. The UN and other organizations frequently deploy teams of Military Observers (UNMOs) to monitor compliance of parties to a truce, accord or international agreement. These personnel are traditionally unarmed. Examples of observer-type missions are UNTSO and ECMMY; and

h. *Humanitarian Assistance Operations*. In addition to normal peacekeeping operations, elements of CF may be called upon by the UN or other international organizations to provide military assistance to humanitarian activities or in aid of disaster relief. Examples are the Mine Awareness and Clearance Training Programmes, or disaster relief operations in Florida.

98   Appendix I

CURRENT TRAINING PROCEDURES

3. CF peacekeepers get the majority of their information and training from three areas: from general CF courses which cover specific peacekeeping topics; from annual refresher training; and from predeployment training. Information and training concerning UN tasks and roles received during CF career courses tends to be general in nature, but essential as a foundation upon which to build. Other military skills, such as a knowledge of Command and Control structures are integral to the success of a peacekeeper in the performance of his duties, but are not identified strictly as peacekeeping skills. The principal reason a good soldier makes a good peacekeeper is the core of general military knowledge and skills he brings to annual refresher and predeployment training sessions.

4. In that no two missions are likely to be faced with identical mandates and circumstances, the best core training for these diverse operations has been found to be general purpose military training, where emphasis is on basic skills and specific to classification skills required by officers and noncommissioned members. To augment these skills, there is a requirement, as with any military operation, for periodic refresher training, specialist training that may be dictated by the mission or circumstance, and comprehensive pre-deployment training.

5. Prior to any deployment, the UN Secretariat normally issues, through the chain of command, training instructions and guidance for the operation. These guidelines are incorporated by NDHQ into training directives and a tasking order to Commands. Following the receipt of this order, the tasked Command normally promulgates its own training direction to units, and they, in turn, review the current state of training and establish priorities for achieving the requisite operational readiness.

6. This training mechanism and procedure utilized by the CF has worked reasonably well in the past, but the lack of a specific organizational structure to direct and monitor the system has made effective standardization and the resulting systemic changes difficult to implement. The present system is not flexible enough to adapt quickly to changing requirements, particularly those driven by evolving mission complexity and it has become apparent that improvements are required.

Appendix I

DETERMINATION OF REQUIREMENTS

7. The process to rectify this training weakness began by developing a summary of UN training requirements as specified in reference C. This list was promulgated to Commands, training establishments and selected NDHQ Directorates at reference D to determine where and to what level peacekeeping skills are developed, and to solicit suggestions for improvements. Additional information regarding the conduct and effectiveness of CF peacekeeping training was also gleaned from the following sources:

   a. critiques of various peacekeeping courses and indoctrination sessions;

   b. after action reports and comments received from missions and Contingent Commanders;

   c. studies of peacekeeping training conducted by other nations; and

   d. a review of the reports from several formal studies of peacekeeping training such as SCONVA and CRS E2/90.

FINDINGS

8. It is the accepted dictum that the CF takes trained soldiers, sailors and airmen and provides specialized training prior to deploying them on a peacekeeping mission. This training has been developed by the Commands and conducted in accordance with UN training guidelines amplified with the experience gained in peacekeeping operations over the last forty years. It concentrates on theatre operations, the mission mandate and environmental, cultural and administrative preparation for individuals or formed units. It is normally decentralized, directed to the specific mission, and its length will vary depending on the urgency of the deployment and the nature of the task. As an adjunct to this training, individuals and units may also conduct acclimatization training and receive up-to-date mission briefings on arrival in theatre.

9. Land Force Command (LFC) has been the primary source of service members for "traditional" peacekeeping operations, except where there

are specific naval or air tasks (eg, UNTAC Coastal/River Patrolling or in-theatre airlift operations and coordination). Generic tasks such as engineering or logistics are not normally element dependent, but if other than Land Force personnel are used to augment composite or formed units or to perform the duties of military observers, then there is a requirement to provide additional refresher and mission specific training. The training for formed units is normally conducted at the designated mounting base by Command-level training personnel augmented as required with guest speakers and specialists. At present UNMO or individual personnel training is normally coordinated by NDHQ training staff, augmented by a specialist training cadre.

10. The documents and recent studies discussed in para 7 have revealed that, while the type and scope of training received is generally very good, there are several critical areas which require additional emphasis and reinforcement, especially mission specific ones, just prior to deployment. Also, although extensive experience exists on peacekeeping training in the field, little formal written guidance exists on course content other than that contained in UN training manuals and locally produced pamphlets. Training SOPs are quite often produced for single missions and are unit specific, thereby rendering them of limited use to other units. Finally, it has become apparent that the CF has a requirement for a visible focal point for peacekeeping training, standardization and doctrine, a so called "center of excellence" to bring all of this expertise and knowledge together.

EXECUTION

11. To ensure that all members of the Canadian Forces who are tasked to participate in peacekeeping operations receive the best possible training and preparation possible, the following improvements and changes will be implemented:

   a. create and establish mechanisms and procedures to ensure:

      (1) an accurate definition of training requirements;

      (2) cost-effective training strategies;

(3) effective management practices for the production, distribution, configuration and control of course-related materials; and

(4) a sound instructional analysis capability; (OPI-J3 Pk)

b. review and update manuals, training packages and aide-memoires of both a general and mission specific nature; (OPI-J3 Pk, OCI-DGRET/DIT)

c. formalize pre-deployment training and packages for individuals, units and formed units; (OPIJ3 Pk, OCI-DGRET/DIT)

d. expand the emergency first aid courses with a view to including training personnel in: cardiopulmonary resuscitation (CPR); injections and administration of painkillers (ie, morphine or demerol), including an explanation and observation of the administering of intravenous (IV) solutions; and the treatment of gunshot or other traumatic wounds (ie, mine, explosive or high explosive shell); (OPI-DMO)

e. review the composition of medical kits issued to military observers or other Canadian Forces personnel that are posted to remote or isolated areas, for new missions, with a view to increasing their contents. For example, antibiotics, intravenous kits, painkillers, needles and syringes, rubber gloves, or other pertinent medical supplies; (OPI-DMO)

f. develop annual refresher training for formed units to place more emphasis on small arms, and where applicable, crew served weapons training, rules of engagement, physical and mental fitness, first aid, driving and operating manual transmission, four wheel drive vehicles (especially for officers assigned to observer missions), battle procedure for living in difficult or extreme field conditions, and the use of personal protective equipment (ie, helmets, body armour, blast blankets, NBCD equipment); (OPI-LFC)

g. formalize pre-deployment training so that it concentrates on mission specific training, an intensive review and confirmation of general military skills personal and personnel administration. Training plans are to be devised so that training can be conducted at a designated mounting base, as close to deployment time as possible. Suggested training requirements are attached as appendices to this annex; (OPIs-J3 Pk, LFC, AIRCOM, MARCOM)

h. review and confirm scales of issue and kit lists for all current missions and potential mission areas based on geographic locations; (OPI-J4 Log)

j. develop training and general information packages for recruit and basic officer training as well as for junior and senior leader courses, the officer professional development programme, staff school and staff colleges. For Officers, these should flow from the Officer General Specification (OGS) and should deal with UN organizations, peacekeeping operations, mounting operations from a Canadian perspective, the UN staff system, UN logistics system and other UN or international agencies; (OPI-DGRET/DIT, OCI-J3 Pk)

k. continue the development of a concept for a peacekeeping centre of excellence by preparing a detailed option analysis for presentation to senior management. The analysis will confirm the requirement for the centre, propose a role and functions and make recommendations for the location, staff, command and control and budget; (OPI-COS J3/DGMPO)

m. create a tailored training package for the reserves, based on regular force training programs, that can be integrated into a regular force program as required; and (OPI-J3 Res, OCI-J3 Pk)

n. develop a training package or course to assist in teaching Canadian Forces personnel the following information:

   (1) information gathering techniques;

   (2) escape and evasion techniques;

   (3) counter-terrorism security measures;

   (4) anti-hijacking techniques;

(5) mediation and negotiation skills; and

(6) how to cope with being captured, conduct during capture or being held hostage. (OPI-J3 Pk).

Appendices:

Appendix 1 UNMO, Staff Officer and Small Composite Unit Pre-Deployment Training Requirements.

Appendix 2 Formed Unit Pre-Deployment Training Requirements

APPENDIX 1 TO ANNEX A
TO 4500-1 (DCDS)
29 DECEMBER 1993

MILITARY OBSERVERS, STAFF OFFICERS AND SMALL COMPOSITE UNITS
PRE-DEPLOYMENT TRAINING REQUIREMENTS

Block 1. – UN Procedures, Missions, Mandates			
Serial	*Subject*	*Material to Be Covered*	*Remarks*
1	Introduction	Opening Remarks, Review Timetable, Distribution of Reference Material and Course Training Package. Administration	
2	United Nations Briefings	UNNY HQ Structure and Mandates. Secretary General and Security Council. Procedures and Process	
3	Legal Briefing	Legal aspects of UN operations. ROEs/Geneva Convention	
4	UN Duties	Duties of a Military Observer and Staff Officer with the UN	

Block 2. – Mission Specific Information			
Serial	*Subject*	*Material to Be Covered*	*Remarks*
5	Peacekeeping	Canadian Peacekeeping Operations Operations – General Overview	
6	Mission Briefing	Mission Specific Operations and Mandate Briefing. UN ORBAT	
7	Geo-Political/ Intelligence Briefing	Mission Area Intelligence. Geography, Cultural, Ethnic, Political Factors	
8	Mission Area Information	Guest Speaker with Mission Area experience. Q&A Session	Current or recently returned Officer.
9	Recognition	Mission Area Equipment Recognition	

Appendix I

		Block 3. - Pers Adm/Medical	
Serial	Subject	Material to Be Covered	Remarks
10	Medical Briefing	Personal Health and Hygiene. Disease prevention. Stress Awareness. Immunizations	
11	First Aid Training	Training/Confirmation of First Aid and CPR Training	
12	Customs Briefing	Canadian Customs regulations. Importing of acquired items.	
13	Media Relations	Contact with the Media. Interview Techniques. Videos	
14	Family Welfare	SISIP Briefing. Will preparation. Pwr of Attorney.	
15	Compensations and Benefits	Pay and Allowance arrangements. Family and personal Travel. Leave entitlements	

		Block 4. - Fieldcraft Skills	
Serial	Subject	Material to Be Covered	Remarks
16	Weapon Training	Personal Weapon Handling, Firing, Evaluation, Cleaning	
17	Driver Training	4 Wheel Drive training. Daily Maintenance. Basic repairs	
18	Personal Equipment	Issue and Confirmation of Kit and Equipment	
19	Survival Skills	Erecting shelters, Field cooking, Water purification	
20	Communications	INMARSAT, VHF/UHF Radios. HF Procedures. Comm Procedures	
21	Mine Awareness Training	Mine Threat, Recognition, Booby traps, Theory, Procedures.	
22	Navigation Training	Use of GPS. Map reading. Compass use.	
23	Personal Training Equipment	Confirmation on Use of Personal Protective Equipment (Helmet, Body Armour, Blast Blankets, and NBCD)	

Block 5. – Operating Techniques/Information

Serial	Subject	Material to Be Covered	Remarks
24	Field Duties	Information Gathering Techniques. Negotiation/Mediation Techniques	
25	Operations	Operating in a Combat Zone. Crossing cease-fire lines	
26	Language Training	Basic Action words to permit communication	Can be provided phonetically on cards
27	Crisis Management	Techniques of self-defence, escape and evasion, anti-hijacking and conduct if captured or held hostage.	
28	Critique	Session Critique.	

APPENDIX 2 TO ANNEX A
TO 4500-1 (DCDS)
29 DECEMBER 1993

FORMED UNIT
PRE-DEPLOYMENT TRAINING REQUIREMENTS

1. Whenever a formed unit is tasked (primarily from Land Force Command), the designated unit normally has a 90 day period prior to deployment to conduct intensive training and preparation. Within this concentrated training period, all of the topics found in the UN Training guidelines are covered.

2. The amount and type of training required for a formed unit is too extensive to be deliniated [sic] here. As a guide, the following general sequence of preparation and training is used:

   a. *Week 1.* Personnel are identified, additional requirements are sourced and received, a departure assistance group is formed and all are assembled at the mounting base. Equipment preparations begin; and

   b. *Week 2.* Stage 1 level individual training is confirmed and conducted as required. Equipment preparations continue;

      (1) Stage 1 - Individual (General)

      (a) general mission briefings;

      (b) Stage 3 Shoot to Live (personal weapons training);

      (c) communications training;

      (d) crew served weapons training;

      (e) individual NBCD training; and

      (f) personal and personnel administration;

108   Appendix I

c. *Week 3 and 4.* Stage 2 level collective training is conducted. Final preparations, maintenance and preparations of vehicles, equipment and unit material are completed:

   (1) Stage 2 - Collective

   (a) Section, platoon and company skills and procedures. This includes mounted operations, road blocks, check points and observation posts, convoy escort and special to mission tactical tasks;

   (b) foot and vehicle patrolling;

   (c) ambush drills;

   (d) crowd control drills;

   (e) use of force and rules of engagement;

   (f) battle tasks as directed by the Commander; and

   (g) searches;

d. *Week 5.* Stage 3 level training is conducted. Vehicles and equipment should be at the Sea Port of Embarkation;

   (1) Stage 3

   (a) local geography, customs, culture(s), politics; (Hand-out booklet)

   (b) intelligence briefings;

   (c) combat first aid;

   (d) how to identify and deal with combat stress;

   (e) demolitions, mines and theatre specific booby traps;

   (f) map reading;

(g) sentry and guard duties;

(h) counter terrorism security measures;

(j) confirmation of live-fire training;

(k) equipment, vehicle and aircraft recognition (booklets required); and

(m) investigations, negotiation/mediation;

e. *Week 6 and 7.* Final pre-embarkation, preparations, finalization of all 3 levels of training confirmed, embarkation leave may be taken, embarkation and movement into the theatre is conducted;

f. *Week 8 and 9.* Acclimatization training begins, equipment arrives, is unloaded and is assembled;

(1) Acclimatization Training

(a) additional training and exercises based on the reconnaissance report and Force Commander's recommendations; and

(b) confirmation of stage training in Canada.

g. *Weeks 10 to 12.* Unit is assigned to operational control of the Force Commander. Acclimatization training and confirmation of all 3 stages of training conducted in Canada is completed.

APPENDIX II

4500-1 (DCDS) 14 Sep 1995 Peacekeeping Training in the Canadian Forces

The Deputy Chief of Defence Staff issued this directive to set in motion a study evaluating peacekeeping training in the Canadian Forces (See Chapter 2).

4500-1 (DCDS)

14 September 1995

Distribution List

DCDS STUDY DIRECTIVE
PEACEKEEPING TRAINING IN THE CANADIAN ARMED FORCES

Refs: A.   4500-1 (DMTE) 14 Jul 95 (NOTAL)
            B.   4500-1 (DCDS) 14 Aug 95 (NOTAL)

1. Enclosed is a copy of the Directive implementing a study on peacekeeping training in the CF. The aim of the study is to evaluate peacekeeping training required in addition to normal combat and occupational training, and to identify the most effective and cost efficient manner of achieving this training. Within these considerations, the study will also identify which training requirements could be best met by the Pearson Peacekeeping Centre (PPC).

2. J3 Trg & NBC will be the OPI for this project. The study will be conducted by CFRETS's training development organization, under the auspices of DGRET/DMTE. The Working Group, as outlined in the enclosed study directive, is responsible for providing overall user guidance and direction leading to the development of a project study plan and for identifying information sources, survey populations, subject matter experts, existing training programmes and related training documents.

3. Based on considerable CF peacekeeping experience in recent years, I believe it is appropriate for us at this time to review again how we train CF personnel for deployment to the various peace support mission areas and to evolve appropriate strategies for our training methods. I urge you to give this study your personal attention and a high staffing priority.

[signed]
Armand Roy
Lieutenant-General

Enclosure: 1

DISTRIBUTION LIST

Action	*Information*
Comd MARCOM	DGRET
Comd LFC	DMTE
Comd AIR COM	DI Pol
Comd CFRETS	J3 Ops
ADM (Per)	J3 Trg & NBC
ADM (Pol & Comm)	
COS J3	

Appendix II

DCDS STUDY DIRECTIVE
WORKING GROUP (WG) ON PEACEKEEPING TRAINING IN
THE CANADIAN FORCES

BACKGROUND

1. The Canadian Institute of Strategic Studies, at the invitation of the Canadian Government, established the Pearson Peacekeeping Centre (PPC), at the former CF Base at Cornwallis. This establishment operates as an independent and private venture, mandated to provide research and education on peacekeeping while serving as a uniquely Canadian point of contact for peacekeeping information. It was created to serve the needs of the entire Peacekeeping Partnership. This term applies to the military, government and non-government agencies dealing with humanitarian assistance, refugees and displaced persons, election monitors, media and civilian police personnel as they work together to improve the effectiveness of peacekeeping operations. In this context, the participation of individuals from all agencies on the courses offered by the PPC is seen as a necessary element in learning from each other's experiences. Accordingly, the CF has provided eight to ten individuals on each of the various PPC courses since they started in the spring of 1995. Although these courses have been well conducted, no analysis of the requirement for them, as they apply to the CF, has been conducted.

AIM

2. The aim of this directive is to detail the WG terms of reference, composition and milestones for a study on CF peacekeeping training (less basic combat and occupational training).

CF PEACEKEEPING TRAINING PHILOSOPHY

3. It is the view of the Canadian Forces that troops assigned to peacekeeping duties need to be well trained in conventional military skills and operations, and that the best core trainer for such duties is general purpose military training, with emphasis on basic combat and specific-to-occupation skills. This, including annual and pre-deployment refresher training in these areas, fulfils the majority of the CF training

requirement for unit and individual participation in UN peacekeeping operations.

4. It is also recognized that there is an additional requirement for a training overlay in UN and Mission specific subjects, such as UN organizations, local history, local culture, local customs and practices, the legal aspects of peacekeeping, current intelligence and operations, conflict resolution, mediation, and negotiation. This training is achieved during:

   a. pre-deployment training for units and sub-units;

   b. the eight day UN Observer and Staff Officer Course as appropriate;

   c. the ten day UNDOF pre-deployment training course as appropriate; and

   d. the three day briefing sessions conducted by NDHQ for personnel posted to UN Missions at short notice (normally individual replacements).

TERMS OF REFERENCE

5. Keeping in mind the CF philosophy on training personnel for peacekeeping duties (paras 3 & 4), the WG is to examine the CF requirement for peacekeeping training, additional to what is achieved through normal combat and occupational training. This will entail a detailed examination of what additional training is required for all CF personnel deploying on peace support operations and an analysis of whether this training requirement is being effectively met by current CF training. In addition, the examination of training is to include special training requirements in specific areas that might be necessary for some of the CF personnel deploying on peace support operations.

6. The WG is to recommend a training strategy that best meets the requirements for conducting additional peacekeeping training. Considering the CF requirement to work with and better understand the other members of the "Peacekeeping Partnership" and the necessity to ensure that all training is economically viable, specific recommendations are to be made concerning what additional training, if any, could

be conducted effectively, on behalf of the CF, by the PPC through existing courses, new courses, or modifications to present courses.

7. Sufficient data to support the validity of the study results, including applicable costs, is to be collected and analysed. Where possible, best use is to be made of the data collected and analysed in the conduct of recent related studies.

WG COMPOSITION

8. The WG will comprise:

   a. Chairman: J3 Trg 2;

   b. Assistant Chairman/Secretary: J3 Trg 2-7;

   c. Training Development Officer: Maj, DMTE staff, NDHQ;

   d. Members:

   (1) LCdr, training staff, MARCOM HQ

   (2) Maj, training staff, LFCHQ

   (3) Maj, training staff, AIRCOM HQ

   (4) Maj, training staff, HQ CFRETS (on behalf of CFCSC and the CNSS);

   (5) Maj, J3 Ops staff, NDHQ;

   (6) Maj, DI Pol staff, NDHQ; and

   (7) Subject matter experts, as deemed appropriate by the WG chairman.

9. Commands and NDHQ staffs are to submit names, appointments and telephone numbers of WG members to the chairman by 20 Sep 95.

10. Meetings are to be convened at the discretion of the WG chairman.

116   Appendix II

FINANCE

11. TD and travel costs associated with this study are to be charged against the Peacekeeping Training Budget. All charges against this budget are to be approved by the budget resource manager, J3 Trg 2-7 (613-995-0852) in advance of any travel.

LIAISON

12. Direct liaison between members of the WG is authorized. In collecting data or arranging for resources in support of the study, the WG is to adhere to the normal chain of command.

MILESTONES

13. Estimating that the study will take up to four months to complete, the following milestones are to be met:

   a. convene initial meeting, by 29 Sep 95;

   b. conduct pre-concept analysis, including target population study, 9-28 Oct 95;

   c. carry out training needs analysis, including collection of data, 30 Oct 95 - 2 Feb 96;

   d. treat, analyse and interpret data, 5 Feb - 1 Mar 96;

   e. write report, including recommendations and business case, 4-29 Mar 96;

   f. submit study report to DCDS, 29 Mar 96.

Appendix II

Memorandum

4500-1 (DMTE)

14 July 1995

Distribution List

TRAINING REQUIREMENT
PEARSON PEACEKEEPING CENTRE (PPC)

Refs: A. 4500-1 (DCDS) 29 Dec 93
B. 1530-1 (DI Pol 6) 7 Jun 95
C. 1530-1 (DI Pol 6) 9 Jun 95
D. 1530-1 (DI Pol 6) 26 Jun 95
E. 3151-9-6 (DCDS) 9 Mar 95

1. Now that the third PPC course has been completed, it is time to reconsider the place of the PPC in the spectrum of training available to the CF. For a variety of reasons, no analysis of the requirement for these courses was conducted prior to the establishment of the PPC.

2. DCDS direction on peacekeeping training, ref A, provides a good yardstick against which to evaluate PPC courses. Although the critiques at ref B through D are good subjective indicators of how the courses at PPC are being conducted, they do not provide an objective analysis of course content.

3. It is clear that we will have a continuing involvement with the PPC at least for the initial few years until it becomes self-supporting. Taking that into account, and in light of recent moves to establish our own peacekeeping training staff, ref E, it behooves us to do a proper requirements and business case analysis for the entire range of peacekeeping training in the CF.

4. As the DCDS group is the OPI for this activity, I would suggest DCDS should take the lead, with ADM(Pol Comm) and my own organization

as OCIs in our areas of expertise. Within ADM(Per), DGRET/DMTE is the point of contact on analysis of training requirements, and will be available to assist you on this important project.

[signed]
PG Addy
LGen
ADM(Per)
992-7582
Fax 995-4519

Appendix II

4500-1 (DCDS)

14 Aug 95

Distribution List

TRAINING REQUIREMENT
PEARSON PEACEKEEPING CENTRE (PPC)

Ref: 4500-1 (DMTE) 14 Jul 95

1. As you have pointed out at the referenced memo, I agree there is a need to complete an analysis of peacekeeping training, with emphasis on identifying specific training requirements that might have to be met using the resources of outside agencies such as the PPC. Accepting the lead in this matter, J3 Trg & NBC has been designated as the OPI.

2. With the assistance of DMTE staff, the J3 Trg & NBC staff will develop a study plan, to include the composition of a suitable study group, the outline objectives and the appropriate time lines. I expect this plan to be complete by the end of Aug 95, at which time I will be able to inform all concerned on the proposed "way ahead".

[signed]
Armand Roy
LGen
992-3395

DISTRIBUTION LIST

Action	*Information*
ADM(Per)	ADM(Pol & Comm)
	DI Pol
	DGRET
	J3 Trg & NBC

4979-8 (DMTE 5-2)

18 September 1995

Distribution List

TRAINING DEVELOPMENT SERVICES PROGRAMME (TDSP)
PEACEKEEPING TRAINING IN THE CANADIAN FORCES

References: A. 4500-1 (DCDS) DCDS Study Directive 14 September 1995 (NOTAL)
B. Telecon Maj Guénard DMTE 5-2/Lcdr Syvertsen-Bitten 7 September 1995
C. Telecon Maj Guénard DMTE 5-2/LCol Reid J3 Trg (4) 7 September 1995

1. It is the view of the Canadian Forces that troops assigned to peacekeeping duties need to be well trained in conventional military skills and operations, and that the best core trainer for such duties is general purpose military training, with emphasis on basic combat and specific-tooccupation skills. This, including annual and pre-deployment refresher training in these areas, fulfils the majority of the CF training requirement for unit and individual participation in UN peacekeeping operations.

2. It is also recognized that there is an additional requirement for a training overlay in UN and Mission specific subjects, such as UN organizations, local history, local culture, local customs and practices, the legal aspects of peacekeeping, current intelligence and operations, conflict resolution, mediation and negotiation. This training is achieved by different means and at different places during the pre-deployment phase of any operation.

3. Keeping in mind the CF philosophy on training personnel for peacekeeping duties, there is a necessity to examine the CF requirement for peacekeeping training, additional to what is achieved through normal combat and occupational training. This will include the following;

   a. the conduct of a detailed needs analysis of what additional training (including special training) is required for all CF personnel deploying on peacekeeping support operations;

121   Appendix II

b. an analysis of whether that training requirement is being effectively met by current CF training; and,

c. a recommendation as to the best training strategy to satisfy those needs.

4. Considering the CF requirement to work with and better understand the other members of the "Peacekeeping Partnership" and the necessity to ensure that all training is economically viable, specific recommendations are to be made concerning what additional training, if any, could be conducted effectively, on behalf of the CF, by the Pearson Peacekeeping Centre (PPC) through existing courses, new courses, or modifications to present courses.

5. In that context, CFRETSHQ is to provide the Training Development Officer support under the TDSP in support of this project. This project is to be assigned a TDSP Priority 1 with a completion date consistent with the DCDS Study Directive enclosed. The final project workplan and schedule of activities are to be determined by the project officer in discussion with the DMTE 5 and J3 Trg project staffs. The project directive, attached at Annex A, and enclosed DCDS Study Directive provides additional details.

[signed]
P.J. Holt
Colonel
Director, Military Training and Education
for Chief of the Defence Staff
(613) 996-6349

Annex:
Annex A - Project directive - Peacekeeping training in the Canadian Forces

Enclosure: 1

DISTRIBUTION LIST

Action	*Information*
CFRETSHQ/DCOS OCC TRG/SSO ITMIS	CFLSTC/Comdt
	J3 Trg NBC
	J3 Trg (2)
	J3 Trg 2-7
	DMTE 4
	DMTE 5

ANNEX A
TO 4979-8 (DMTE 5-2)
DATED 14 SEPTEMBER 1995

TRAINING DEVELOPMENT SERVICES PROGRAMME (TDSP)PROJECT DIRECTIVE
PEACEKEEPING TRAINING IN THE CANADIAN FORCES

References: A. 4500-1 (DMTE) 14 July 1995
B. 4500-1 (DCDS) 14 August 1995
C. DCDS Study Directive 14 September 1995 (enclosed)
D. Telecon Maj Guénard DMTE 5-2/LCdr Syvertsen-Bitten CFLSTC 7 September 1995
E. Fax CFLSTC 7 September 1995

BACKGROUND

1. The Canadian Institute of Strategic Studies, at the invitation of the Canadian Government, established the Pearson Peacekeeping Centre (PPC), at the former CF Base of Cornwallis. This establishment operates as an independent and private venture, mandated to provide research and education on peacekeeping information. It was created to serve the needs of the entire Peacekeeping Partnership. This term applies to the military, government and non-government agencies dealing with humanitarian assistance, refugees and displaced persons, election monitors and media, and civilian police personnel as they work together to improve the effectiveness of peacekeeping operations. In this context, the participation of individuals from all agencies on the courses offered by the PPC is seen as a necessary element in learning from each other's experiences. Accordingly, the CF has provided eight to ten individuals on each of the various PPC courses since they started in the spring of 1995. Although these courses have been well conducted, no analysis of the requirement for them, as they apply to the CF, has been conducted.

PROJECT AIMS

2. The aims of this project are:

Appendix II

a. to conduct a detailed needs analysis of what additional training (including special training) is required for all CF personnel deploying on peacekeeping support operations;

b. to conduct an analysis of whether that training requirement is being effectively met by current CF training;

c. to recommend the best training strategy to satisfy the prioritized needs identified by the target population; and,

d. to recommend what additional training, if any, could be conducted effectively, on behalf of the CF, by the PPC through existing courses, new courses, or modifications to present courses.

SCOPE

3. The project scope to achieve the aims should include the following activities:

a. liaise with DMTE & J3 staffs to ensure that the project officer(s) will adopt an adequate needs analysis model as well as the most appropriate tools for data gathering;

b. develop a project plan and schedule that will meet the milestones set in the DCDS directive;

c. define the problematic situation including limits, variables and the target population;

d. gather and validate data from subject matter experts (SMEs) identified by the J3 point of contact;

e. adoption, adaption/creation, validation and administration of a data gathering tool which must be administered to the target population;

f. treatment and prioritization of needs;

g. put aside needs that are currently satisfied by CF training establishments;

h. make a recommendation on needs identified at 3g above;

i. produce a list of additional training needs not currently being looked after by CF training establishments;

j. recommend the most efficient training strategy for the additional training needs that have been prioritized and are not being satisfied by CF training establishments;

k. recommend what additional training, if any, could be conducted effectively, on behalf of the CF, by the PPC through existing courses, new courses, or modifications to present courses;

l. asses[s] costs and benefits of the PPC option;

m. asses[s] feasibility and risks of implementing the PPC option;

n. prepare a project report to include:

   (1) activities conducted and results; as well as,

   (2) recommendations for course of action based on project activities; and,

o. because of the nation wide implication of the study, the report must be reviewed by DMTE prior to final submission to the client.

TIMINGS

4. The DCDS study directive (enclosed) outlines the time estimates as well as the milestones for the project.

ADMINISTRATION

5. *Project Sponsor.* J3 Trg/NBC shall be the sponsor for this project.

6. *NDHQ Points of Contact* (POC). Maj G. Currie, J3 Trg 2-7, at (613) 995-0852 will be the principal NDHQ POC and act as a (SME). Maj C. Guénard, DMTE 5-2, at (613) 995-8303 will be the secondary NDHQ POC.

7. *Funding.* Subject to the requirements of paragraph 10, J3 Trg special project office will arrange for funding for any appropriate temporary duty associated with this specific TDSP project. All charges against this budget are to be approved by the budget resource manager, J3 Trg 2-7 in advance of any travel. Copies of finalized claims shall be forwarded to J3 Trg 2-7 for audit purposes.

8. *Lines of Communication.* Direct liaison is authorized between the CFLSTC Project Officer(s) and NDHQ POC after formal tasking by CFRETSHQ.

9. *Reports.* A final report addressing the requirements of paragraphs 2 and 3 shall be forwarded for review by DMTE 5-2 prior to final submission to J3 Trg/NBC.

SCHEDULE

10. A project workplan, including the proposed budget, shall be forwarded to the J3 Trg POC for review within 3 weeks of project tasking. As discussed at refs D and E, project planning should start on 9th October 95 and TDSP Project Report (with business case) should be submitted on 29th March 1996. Remaining project activities shall be initiated as per the approved workplan.